AF347525

Innovative Preservation Technology for the Fresh Fruit and Vegetables

Innovative Preservation Technology for the Fresh Fruit and Vegetables

Editors

Bernardo Pace
Maria Cefola

MDPI • Basel • Beijing • Wuhan • Barcelona • Belgrade • Manchester • Tokyo • Cluj • Tianjin

Editors
Bernardo Pace
Institute of Sciences of Food
Production
CNR—National Research
Council of Italy
Foggia
Italy

Maria Cefola
Institute of Sciences of Food
Production
CNR—National Research
Council of Italy
Foggia
Italy

Editorial Office
MDPI
St. Alban-Anlage 66
4052 Basel, Switzerland

This is a reprint of articles from the Special Issue published online in the open access journal *Foods* (ISSN 2304-8158) (available at: www.mdpi.com/journal/foods/special_issues/Innovative_Preservation_Technology_Fresh_Fruit_Vegetables).

For citation purposes, cite each article independently as indicated on the article page online and as indicated below:

LastName, A.A.; LastName, B.B.; LastName, C.C. Article Title. *Journal Name* **Year**, *Volume Number*, Page Range.

ISBN 978-3-0365-1330-0 (Hbk)
ISBN 978-3-0365-1329-4 (PDF)

© 2021 by the authors. Articles in this book are Open Access and distributed under the Creative Commons Attribution (CC BY) license, which allows users to download, copy and build upon published articles, as long as the author and publisher are properly credited, which ensures maximum dissemination and a wider impact of our publications.

The book as a whole is distributed by MDPI under the terms and conditions of the Creative Commons license CC BY-NC-ND.

Contents

About the Editors

Bernardo Pace

Bernardo Pace is a researcher at the ISPA—CNR, with significant experience in pre- and post-harvest of fruits and vegetables in international and national projects. He participated in cooperation projects in Tunisia and Albania and in many international courses and congress on post-harvest activities. He spent a period at Mann Lab of UC Davis, expanding on post-harvest research. As a researcher, he assumed the role of project leader and led the research activity and has been involved in activities in the follow thematic areas: identification of technologies to improve the quality of products, whole and fresh-cut applications; innovative treatments and/or pre-treatments; new packing materials and storage conditions; valorization of by-products of fruit and vegetables, and application of non-destructive systems able to assess the quality of fruit and vegetables. He is the author of more than 80 scientific papers and book chapters.

Maria Cefola

Maria Cefola is a researcher at the ISPA—CNR. From 2006 to 2010, she attended the Ph.D. School in Innovation Management in Agri-Food Systems in the Mediterranean, researching post-harvest management of fresh and fresh-cut fruit and vegetable products of the Mediterranean region at the University of Foggia. During her doctorate, she spent six months at the Department of Plant Science of UC Davis. In 2004, she obtained a Master's Degree in Food Science and Technology at the University of Basilicata. She has had significant experience in post-harvest management of fruits and vegetables within international, national and regional research projects and has attended international courses and congress. Her research activities regard: the improvement of the quality of fresh-cut products through the application of innovative pre-treatment, packaging and/or storage conditions; innovation in logistic cold chain; the study of innovative non-destructive systems for quality evaluation.

Editorial

Innovative Preservation Technology for the Fresh Fruit and Vegetables

Bernardo Pace * and **Maria Cefola**

Institute of Sciences of Food Production, National Research Council of Italy (CNR), c/o CS-DAT, Via M. Protano, 71121 Foggia, Italy; maria.cefola@ispa.cnr.it
* Correspondence: bernardo.pace@ispa.cnr.it

Abstract: The preservation of the freshness of fruits and vegetables until their consumption is the aim of many research activities. Quality losses of fresh fruit and vegetables during cold chain are frequently attributable to an inappropriate use of postharvest technologies. Moreover, especially when fresh produce is transported to distant markets, it is necessary to adopt proper postharvest preservation technologies in order to preserve the initial quality and limit microbial decay. Nowadays, for each step of supply chain (packing house, cold storage rooms, precooling center, refrigerate transport and distribution), are available innovative preservation technologies that, alone or in combination, could improve the fresh products in order to maintain the principal quality and nutritional characteristics. The issue groups five original studies and two comprehensive reviews within the topic of preservation technologies related to innovative packaging and postharvest operation and treatments, highlighting their effect on quality keeping.

Keywords: active cardboard box; antimicrobial compounds; biocontrol; nanoparticles coating; oxalic acid; plasma-activated water; preservatives; respiration rate; shelf life; TiO_2 photocatalytic

check for **updates**

Citation: Pace, B.; Cefola, M. Innovative Preservation Technology for the Fresh Fruit and Vegetables. *Foods* **2021**, *10*, 719. https://doi.org/10.3390/foods10040719

Received: 23 March 2021
Accepted: 27 March 2021
Published: 29 March 2021

Publisher's Note: MDPI stays neutral with regard to jurisdictional claims in published maps and institutional affiliations.

Copyright: © 2021 by the authors. Licensee MDPI, Basel, Switzerland. This article is an open access article distributed under the terms and conditions of the Creative Commons Attribution (CC BY) license (https://creativecommons.org/licenses/by/4.0/).

Fresh fruits and vegetables are perishable food, which undergo quality losses during cold chain if suitable preservation technologies are not used. Among these, packaging has an important role. Recently, active packaging was successful applied to improve the shelf life of fresh table grape [1] and blueberries [2]. Furthermore, active packaging, using red thyme oil (*Thymus vulgaris* L.), could be employed by the citrus industry to extend the shelf-life of oranges for fresh market use and juice processing [3].

Moreover, postharvest processes and technologies may preserve the quality and limit microbial decay in fresh fruits and vegetables. In this context, Pinto et al. [4] showed that the exposure to gaseous ozone before packaging followed by storage under modified atmosphere packaging (MAP) could be a useful technological approach to extend the postharvest storage of small berry fruit. Similarly, the combined effect of cold storage and oxalic acid treatment resulted in a valid and sustainable solution to preserve the visual quality of green and purple asparagus spears [5].

In the Special Issue on "Innovative Preservation Technology for the Fresh Fruit and Vegetables", three main research topics were covered: (i) innovative packaging; (ii) postharvest processes and technology affecting the product quality; (iii) postharvest technology to limit microbial decay. Within the first topic, three research articles were published.

In the first article [6], the use of cardboard box activated with essential oils in order to enhance the shelf life of fresh mandarins (*Citrus reticulata* Murcott Seedless) was described. The authors studied the effect of different forms of paperboard packages sprayed in internal surfaces with the lacquer containing essential oils (EOs) encapsulated by cyclodextrins inclusion complex in order to simulate the transport boxes of 10 kg of mandarins fruits until three weeks at 8 °C. This results in reduced microbial growth (together with decay incidence) and weight losses, while maintaining the physicochemical quality (soluble solids, titratable acidity, firmness and color) with an increase of one week in storage life. The

1

controlled release of EOs from the box extended the shelf life of mandarins during either a long cold transportation simulation or a commercialization period at non-recommended room temperature.

In the second paper, Pace et al. [7] studied the proper oxalic acid (OA) concentration and the suitable packaging material able to maintain an adequate low O_2 concentration inside fresh-cut iceberg lettuce (*Lactuca sativa* L.) bags closed in MAP. The results showed a significant effect of 5 mM OA on respiration rate delay. In addition, polypropylene/polyamide (PP/PA) was selected as the most suitable packaging material to be used in low O_2 MAP. Combining OA dipping with low O_2 MAP using PP/PA as material resulted in the ability to reduce leaf edge browning, respiration rate, weight loss and electrolyte leakage, preserving the visual quality of fresh-cut lettuce until 8 days at 8 °C.

The last article, within the first topic, aimed to identify the impact of 15% Kelulut honey (KH) nanoparticles (Nps) coating solution on papaya's respiration rate, antioxidant activity and total phenol content and to investigate the respiration rate kinetic model of nano-coated papaya (*Carica papaya* L.) using the Peleg model in order to describe the function of gas composition and storage day [8]. The results showed that KH Nps coating can be used as a conserving material, extending the shelf life by inhibiting the respiration rate and C_2H_4 production, while maintaining the antioxidant activity and total phenol content, in papaya.

The research topic regarding postharvest processes and technology affecting the product quality was also covered with two research articles published in this Special Issue.

The effect of cutting styles (slice, pie and shred) on the quality characteristics and antioxidant activity of purple and yellow flesh sweet potato cultivars during 6 days of storage at 4 °C was investigated by Dovene et al. [9]. The finding of this study revealed that pie-cut processing has potential in improving the quality and increasing the antioxidant activity of fresh-cut purple and yellow flesh sweet potato (*Ipomoea batatas* L.) cultivars, while shredding accelerated the quality deterioration of both sweet potato cultivars. Jia et al. [10] proposed a precise temperature control cold storage with low-temperature fluctuation (LFT) combined with an ozone (O_3) generator and a titanium dioxide (TiO_2) photocatalytic reactor to the cold storage of peach (*Prunus persica* L. Batsch). The results showed that LFT significantly reduced the chilling injury of peach fruit during storage. Moreover, its combination with the TiO_2 photocatalytic system significantly improved the postharvest storage quality of the fruit. This treatment maintained higher titratable acidity, total soluble solids, better firmness, color, microstructure and lower decay rate, polyphenol oxidase activities, total phenol accumulation, respiratory intensity, ethylene production and malondialdehyde content during 60 d of storage.

Finally, the last research topic was the object of two reviews. In the first one, the recent physical, chemical and the biological approaches conceived to control the development of gray mold *Botrytis cinerea* in table grapes was discussed by De Simone et al. [11]. Since the global consumption of table grape (*Vitis vinifera* L.) has increased in the last 20 years by over 70% [12], the researchers studied solutions to control *Botrytis cinerea*, which represent the major cause of table grapes losses occurring in pre and postharvest. Among physical methods, dipping in hot water, electrolyzed oxidizing water or different gas compositions using controlled or modified atmosphere packaging are discussed. Regarding the chemical methods, different treatments (wound inoculation, spraying, dipping or fumigation) were reported to control *Botrytis cinerea*. In regard to the bio-based applications, several protective cultures and compounds of biological origin, were assessed for their possible use as biological control agents against gray mold decay. Many biological compounds were tested for the biocontrol of table grape spoilages and these compounds include vegetal extracts, essential oils or edible coating. The authors highlight that each treatment has peculiar benefits and limitations that affect the concrete applications and the future perspectives. As previously applied with success in other fields, an integrated management program that contemplate the combinations of two or more different solutions, could be useful to minimize post-harvest losses caused by undesired fungal development on table grape.

The second review elaborated the properties of plasma-activated water (PAW), the effect of various treatment parameters on its efficiency in bacterial inactivation and its usage as a standalone technology, as well as a hurdle approach with mild thermal treatments [13]. A section highlighting different models that can be employed to generate PAW alongside a direct comparison of the PAW characteristics on the inactivation potential and the existing research gaps are also included. The mechanism of action of PAW on the bacterial cells and any reported effects on the sensory qualities and shelf life of food has been evaluated. It was concluded that PAW offers a significant potential as a non-chemical and non-thermal intervention for bacterial inactivation, especially on food. However, the applicability and usage of PAW depend on the effect of environmental and bacterial strain-based conditions and cost-effectiveness.

In conclusion, the research papers proposed, carried out within the European or National Project or promoted by private enterprise, reported innovative results improving the knowledge in the topic of "Innovative Preservation Technology for the Fresh Fruit and Vegetables". We expect the results presented in this Special Issue to be a stimulus for other future research in this field.

Author Contributions: B.P. and M.C. equally contributed to organizing the Special Issue, to editorial work and to writing this editorial. All authors have read and agreed to the published version of the manuscript.

Funding: This research received no external funding.

Institutional Review Board Statement: Not applicable.

Informed Consent Statement: Not applicable.

Data Availability Statement: Not applicable.

Acknowledgments: We thank all authors for submitting the manuscripts of high quality and the reviewers willingness for their careful evaluations in order to improve the papers. All Editorial staff of MDPI for the professional support and the rapid actions promoted throughout the editorial process and the Foods Editorial Office for the opportunity to cover the role of the Guest Editors of this Special Issue.

Conflicts of Interest: Declare conflicts of interest.

References

1. Gorrasi, G.; Bugatti, V.; Vertuccio, L.; Vittoria, V.; Pace, B.; Cefola, M.; Quintieri, L.; Bernardo, P.; Clarizia, G. Active packaging for table grapes: Evaluation of antimicrobial performances of packaging for shelf life of the grapes under thermal stress. *Food Packag. Shelf Life* **2020**, *25*, 100545. [CrossRef]
2. Bugatti, V.; Cefola, M.; Montemurro, N.; Palumbo, M.; Quintieri, L.; Pace, B.; Gorrasi, G. Combined Effect of Active Packaging of Polyethylene Filled with a Nano-Carrier of Salicylate and Modified Atmosphere to Improve the Shelf Life of Fresh Blueberries. *Nanomaterials* **2020**, *10*, 2513. [CrossRef] [PubMed]
3. Pinto, L.; Cefola, M.; Bonifacio, M.A.; Cometa, S.; Bocchino, C.; Pace, B.; De Giglio, E.; Palumbo, M.; Sada, A.; Logrieco, A.F.; et al. Effect of red thyme oil (*Thymus vulgaris* L.) vapours on fungal decay, quality parameters and shelf-life of oranges during cold storage. *Food Chem.* **2021**, *336*, 127590. [CrossRef] [PubMed]
4. Pinto, L.; Palma, A.; Cefola, M.; Pace, B.; D'Aquino, S.; Carboni, C.; Baruzzi, F. Effect of modified atmosphere packaging (MAP) and gaseous ozone pre-packaging treatment on the physico-chemical, microbiological, and sensory quality of small fruit. *Food Packag. Shelf Life* **2020**, *26*, 100573. [CrossRef]
5. Cefola, M.; Pace, B.; Azara, E.; Spissu, Y.; Serra, P.R.; Logrieco, A.F.; D'hallewin, G.; Fadda, A. Postharvest application of oxalic acid to preserve overall appearance and nutritional quality of fresh-cut green and purple asparagus during cold storage: A combined electrochemical and mass-spectrometry analysis approach. *Postharvest Biol. Technol.* **2019**, *148*, 158–167. [CrossRef]
6. López-Gómez, A.; Ros-Chumillas, M.; Buendía-Moreno, L.; Navarro-Segura, L.; Martínez-Hernández, G.B. Active Cardboard Box with Smart Internal Lining Based on Encapsulated Essential Oils for Enhancing the Shelf Life of Fresh Mandarins. *Foods* **2020**, *9*, 590. [CrossRef] [PubMed]
7. Pace, B.; Capotorto, I.; Palumbo, M.; Pelosi, S.; Cefola, M. Combined Effect of Dipping in Oxalic or in Citric Acid and Low O_2 Modified Atmosphere, to Preserve the Quality of Fresh-Cut Lettuce during Storage. *Foods* **2020**, *9*, 988. [CrossRef]

8. Maringgal, B.; Hashim, N.; Mohamed Amin Tawakkal, I.S.; Mohamed, M.T.M.; Hamzah, M.H.; Mohd Ali, M. Effect of Kelulut Honey Nanoparticles Coating on the Changes of Respiration Rate, Ascorbic Acid, and Total Phenolic Content of Papaya (*Carica papaya* L.) during Cold Storage. *Foods* **2021**, *10*, 432. [CrossRef] [PubMed]
9. Dovene, A.K.; Wang, L.; Bokhary, S.U.F.; Madebo, M.P.; Zheng, Y.; Jin, P. Effect of Cutting Styles on Quality and Antioxidant Activity of Stored Fresh-Cut Sweet Potato (*Ipomoea batatas* L.) Cultivars. *Foods* **2019**, *8*, 674. [CrossRef] [PubMed]
10. Jia, X.; Li, J.; Du, M.; Zhao, Z.; Song, J.; Yang, W.; Zheng, Y.; Chen, L.; Li, X. Combination of Low Fluctuation of Temperature with TiO_2 Photocatalytic/Ozone for the Quality Maintenance of Postharvest Peach. *Foods* **2020**, *9*, 234. [CrossRef]
11. De Simone, N.; Pace, B.; Grieco, F.; Chimienti, M.; Tyibilika, V.; Santoro, V.; Capozzi, V.; Colelli, G.; Spano, G.; Russo, P. *Botrytis cinerea* and Table Grapes: A Review of the Main Physical, Chemical, and Bio-Based Control Treatments in Post-Harvest. *Foods* **2020**, *9*, 1138. [CrossRef] [PubMed]
12. FAO; OIV. Table and Dried Grapes: World Data. Available online: http://www.fao.org/documents/card/en/c/709ef071-6082-4434-91bf-4bc5b01380c6/ (accessed on 18 March 2021).
13. Soni, A.; Choi, J.; Brightwell, G. Plasma-Activated Water (PAW) as a Disinfection Technology for Bacterial Inactivation with a Focus on Fruit and Vegetables. *Foods* **2021**, *10*, 166. [CrossRef] [PubMed]

Article

Active Cardboard Box with Smart Internal Lining Based on Encapsulated Essential Oils for Enhancing the Shelf Life of Fresh Mandarins

Antonio López-Gómez [1,2,*], María Ros-Chumillas [1,2], Laura Buendía-Moreno [1], Laura Navarro-Segura [1] and Ginés Benito Martínez-Hernández [1,2]

[1] Food Safety and Refrigeration Engineering Group, Department of Agricultural Engineering, Universidad Politécnica de Cartagena, Paseo Alfonso XIII, 48, 30203 Cartagena, Spain; may.ros@upct.es (M.R.-C.); laura.buendia.moreno@gmail.com (L.B.-M.); laura.navarro@upct.es (L.N.-S.); GinesBenito.Martinez@upct.es (G.B.M.-H.)

[2] Biotechnological Processes Technology and Engineering Lab, Instituto de Biotecnología Vegetal, Universidad Politécnica de Cartagena, Edif I+D+I, Campus Muralla del Mar, 30202 Cartagena, Murcia, Spain

[*] Correspondence: antonio.lopez@upct.es; Tel.: +34-968-325516

Received: 11 April 2020; Accepted: 1 May 2020; Published: 6 May 2020

Abstract: Mandarins are usually sold in bulk and refrigerated in open cardboard boxes with a relatively short shelf-life (12–15 days) due to physiological and pathological disorders (rot, dehydration, internal breakdown, etc.). The influence of a controlled release of essential oils (EOs) from an active packaging (including β-cyclodextrin-EOs inclusion complex) was studied on the mandarin quality stability, comparing different sized cardboard trays and boxes, either non-active or active, at the pilot plant scale (experiment 1; commercialization simulation at room temperature after a previous simulation of short transportation/storage of 5 days at 8 °C). Then, the selected package was further validated at the industrial scale (experiment 2; cold storage at 8 °C up to 21 days). Among package types, the active large box ($\approx$10 kg fruit per box) better maintained the mandarin quality, extending the shelf life from two weeks (non-active large box) to three weeks at room temperature. Particularly, the active large box highly controlled microbial growth (up to two log units), reduced weight losses (by 1.6-fold), reduced acidity, and increased soluble solids (highly appreciated in sensory analyses), while it minimized colour and controlled firmness changes after three weeks. Such trends were also observed during the validation experiment, extending the shelf life (based on sensory quality) from 14 to at least 21 days. In conclusion, the mandarin's shelf life with this active cardboard box format was extended more than one week at 8 °C.

Keywords: β-cyclodextrin; inclusion complex; carvacrol; essential oils; active packaging; citrus; quality; shelf life; decay incidence

---

1. Introduction

Mandarin (*Citrus reticulata*) is a citrus fruit consumed worldwide with a total production of approximately 34 million tonnes in 2018 [1]. The production of mandarins (including clementines, satsumas and different hybrid mandarins) occupies the second place of total citrus fruit production followed by lemons, limes, and grapefruits [1]. Nevertheless, since a high proportion of oranges, the most cultivated citrus fruit, are used for juice extraction [2], mandarin production could thus be considered as the most cultivated citrus fruit for fresh consumption.

The high popularity of mandarins among consumers is explained by its characteristic convenience (easy to peel and eat), palatable sweetness-acidity binomial, and high content of bioactive compounds, mainly vitamin C [3,4]. The sweetness and acidity of mandarins, and citrus fruit in general, have been conventionally measured with soluble solids content (SSC) and titratable acidity (TA) determinations. This important sweetness-acidity binomial has been quantified using maturity indexes like the conventional SSC/TA index, and the BrimA (an abbreviation for "Brix minus Acids") index [5], as recommended for citrus [6]. These maturity indexes have been highly correlated to the fruit flavour together with peel colour, which may influence the sensory acceptance highly influenced by them [6,7]. The development of citrus peel colour during maturation is associated to the coordinated carotenoid biosynthesis and chlorophyll degradation [7]. In particular, up to 94% of total carotenoid content in mandarins are present in the peel after colour beak. Specifically, β-cryptoxanthin and apocarotenoids are the carotenoids responsible for the distinctive orange-reddish pigmentation in mandarins [7]. The peel colour is the main attribute of mandarins determining the consumer decision to purchase them, together with size and blemish-free [8]. Several colour indexes calculated using CIE and Hunter coordinates (mainly L, a and b) have been proposed to determine the orange colour changes in citrus with a single and representative colour index. In particular, *a/b* index, which increases with increasing yellow to red colour, was proposed early for citrus due to the high correlation with the United States Department of Agriculture (USDA) colour standards [9,10]. Furthermore, $((1000 \times a)/(L \times b))$ index was proposed for predicting the fruit colour changes during the degreening of citrus since it correlated well with the visual appreciation of peel colour changes of citrus from dark green to orange [11]. Citrus fruit flesh is not as firm as other fruits due to the characteristic presence of juice sacs. Nevertheless, citrus firmness is important since flesh firmness of citrus fruits, and mandarin in particular, highly influences their mouth feel [12].

Mandarin postharvest losses can have physiological, pathological or physical (e.g., rind wounds, bruises) origin, with weight loss and physiological disorders being the major causes of such postharvest losses [2]. Weight loss does not result only in direct quantitative losses (=economic losses), leading also to the triggering of physiological disorders like peel pitting, stem-end rind breakdown, shrivelling, collapse of the stem-end button, etc. [2,13]. Mandarins are also sensitive to chilling injuries when the storage temperature is below 5 °C, characterized by pitting and brown discolouration followed by increased susceptibility to decay. Therefore, the recommended storage temperature for mandarins is 5–8 °C [13]. Among the main pathological disorders of citrus fruits, mainly caused by filamentous fungi are stem-end rots, green mould, blue mould, grey mould, black rot, brown rot, and anthracnose [2].

Postharvest treatments conventionally used to maintain the quality of citrus and extend their shelf life are waxes and/or chemical fungicides. Nevertheless, such conventional postharvest treatments have raised important health and environmental issues, as the residue levels of such agrochemicals that are progressively more restricted by official regulations. In that sense, alternative and eco-sustainable postharvest treatments are needed, presenting essential oils (EOs) an interesting opportunity due to their high antimicrobial properties.

EOs are oily liquids extracted from plants that display a high in vitro antimicrobial activity. Carvacrol is the major component of oregano EOs with wide spectra against both gram-negative (e.g., enterobacteria) and gram-positive bacteria, and other microbial groups like moulds [14–17]. Nevertheless, in vivo effectiveness of EOs is reduced due to their evaporation and other light and oxygen degradative reactions. In that sense, higher EO concentrations are needed in vivo to reach the same effectiveness in vitro, with the consequent appearance of EO-related off-flavours [18]. Furthermore, EO mixes including the major EO components (e.g., carvacrol) together with its correspondent EOs (e.g., oregano EOs) have shown a synergistic effect on the antimicrobial activity [19]. In that sense, an encapsulated EOs mix composed of carvacrol:oregano EO:cinnamon EO (70:10:20; weight (*w*):*w*:*w*) showed a high antimicrobial effect in plant products packaged with this active package [11,20]. Nanoencapsulation can greatly reduce EOs oxidation and evaporation while ensuring a controlled release of EOs in small concentrations to the surrounding atmosphere. Cyclodextrins (CDs) (cyclic oligomers of α-D-glucopyranose with a hydrophobic cavity) can highly encapsulate EOs avoiding their oxidation, light-degradation, evaporation, etc. [15,21]. The most important CDs at the industrial level are α- and β-CDs. In particular, β-CD is highly extended due to its low cost. βCD is approved as a food additive in Europe (E459), USA, and Japan, with an acceptable daily intake of 5 mg kg^{-1} (body weight) day^{-1} [22].

Antimicrobial active packaging is an emerging technology that allows extending the food shelf life through a controlled release of the encapsulated antimicrobial compounds [23]. Corrugated cardboard is widely used in the European Union as an eco-sustainable packaging material for packaging of fresh fruit and vegetables. Furthermore, cardboard is highly used for mandarins under different formats (boxes, trays, flock-packs, alveoli trays, etc.). In that sense, EOs-βCD inclusion complexes may be included to develop antimicrobial active cardboard packaging to extend the shelf life of plant products as observed by our Group in vegetables [15,20,24,25]. High relative humidity (RH) and temperatures have been reported to increase the controlled release of EOs from the inclusion complex [20,26], as is expected to occur with the recommended high RH (90%–95%) maintained during cold storage of mandarins and subsequent commercialization during retail at room temperature. Nevertheless, the effects of this antimicrobial active packaging have not been studied yet on fruits with a high potential to increase the shelf life of products usually sold in cardboard packages like mandarins.

This work aimed to study the effect of an active (supplemented with an EOs mix-βCDs inclusion complex) cardboard packaging in different formats (different sized trays and boxes) on the mandarin quality after a commercialization simulation (room temperature) at the pilot plant scale up to three weeks (with a previous simulation of short transportation/storage of five days at 8 °C). The selected package was then validated at the industrial scale in a second experiment during storage at 8 °C up to 21 days.

2. Materials and Methods

2.1. Materials

Carvacrol, oregano, and cinnamon EOs were obtained from Lluch Essence S.L. (Barcelona, Spain). βCD (Kleptose®10) was obtained from Roquette (Lestrem, France). Waterproof lacquer (UKAPHOB HR 530; 10.5% solids) (authorised for food contact surfaces in accordance with EC (2004) [27]) was acquired from Schill+seilacher GMBH (Böblingen, Germany). Corrugated cardboard was supplied by SAECO company (Molina de Segura, Spain). All microbial analysis materials were acquired from Scharlau Chemie (Barcelona, Spain).

Mandarins (*Citrus reticulata* Murcott Seedless) were obtained from the company Blancasol S.a.t. (Blanca, Murcia, Spain) in March 2019. Mandarins were grown in open fields under organic agriculture practices. Fruits were manually harvested and transported to the pilot plant of our Department, where they were selected according to homogeneous orange colour (more than 80 percent of the surface showed orange colour), physical integrity and absence of decay. No waxes neither fungicides were applied to mandarins. Selected mandarins were then packaged within the corresponding packaging treatment and then stored as described in each experiment (pilot plant scale and industrial validation experiments, as described below).

2.2. Preparation of the EOs-βCD Inclusion Complex and Application to Packages

An EOs mix composed of carvacrol:oregano EO:cinnamon EO 70:10:20 (weight (*w*):*w*:*w*) (composition analysis previously reported [25]) was prepared based on its high antimicrobial effect in accordance with our previous studies with several EOs mixes [15,28]. The EOs-βCD inclusion complex was prepared using the kneading method [29]. Briefly, 0.15 g of EOs was mixed with 1.14 g of βCD (1:1 molar ratio) in a mortar with 3 mL of ethanol, kneaded for 45 min and finally maintained in a vacuum desiccator at room temperature for at least 72 h. The achieved encapsulation efficiency of the EOs-βCD inclusion was of 92%–95%, in accordance with our previous data [15]. This EOs-βCD inclusion complex has been fully characterized and published by our group [15,25].

The EOs-βCD inclusion complex was dissolved in water-diluted lacquer prior to spraying on all internal surface of the package. The lacquer was diluted (to a final solid concentration of 8.5%) to compensate for the addition of EOs-βCD inclusion complex since lacquers with solid content >30% make lacquer spraying on the cardboard surface difficult. In that sense, the EOs-βCD inclusion complex content was at that maximum concentration that did not compromise the technological properties of the lacquer to be sprayed on the cardboard surface of the packaging in accordance with preliminary tests. Lacquer containing the EOs-βCD inclusion complex was sprayed at 12 mL·m^{-2} following the manufacturer's recommendations to obtain homogeneous spraying on the paperboard surface while reaching a maximum lacquer absorption. The mechanical and hydrophobic properties of the paperboard material, sprayed and non-sprayed, with the EOs-βCD inclusion complex are fully described in our previous publication [15].

2.3. Design of the Pilot Plant Experiment and Industrial Validation Experiment: Packaging Treatments and Storage Conditions

Two experiments were performed during this study: pilot plant scale experiment and industrial validation experiment.

2.3.1. Experiment 1: Pilot Plant Scale

The first experiment was developed at pilot plant scale studying five different packaging formats to select the most appropriate in order to highly maintain the mandarin quality during a short cold storage period (simulating a short transport/storage of 5 days at 8 °C) supplemented with a commercialization period at room temperature up to 3 weeks. The five package formats (Figure 1) were:

- Small tray (ST): Tray (230 × 120 × 25 mm; length × width × height) of microcorrugated paperboard with flow-pack of macroperforated (45 holes (8 mm Ø each) per m^2) film of polylactic acid (PLA). Each ST contained 0.7–0.8 kg of fruit.
- Small box (SB): Box with lid (190 × 150 × 75 mm) of macrocorrugated paperboard. Each lateral side of the box had 2 holes (10 mm Ø each) and the lid 4 holes (10 mm Ø each). Each SB contained 0.7–0.8 kg of fruit.
- Large tray (LT): Tray (300 × 200 × 90 mm) of macrocorrugated paperboard with a cover of the macroperforated PLA film. Each LT contained 2.3–2.4 kg of fruit.
- Large tray+alveoli (LT+): LT with alveoli (also known as pulp tray) of microcorrugated paperboard. Each LT+ contained 8 mandarins.
- Large box (LB): Box (600 × 400 × 180 mm) of macrocorrugated paperboard with a cover of the macroperforated PLA film. Each LB contained ≈10 kg of fruit.

Each of the five package types was also tested as active or non-active packages. For active packages, all internal surfaces of paperboard packages were sprayed with the lacquer containing the EOs-βCD inclusion complex (as described in the previous section). For the active LT+, only the alveoli tray was sprayed with the lacquer containing the EOs-βCD inclusion complex. For non-active packages, the same procedure was conducted but the lacquer did not include the EOs-βCD inclusion complex.

All packages with fruit were stored in this first experiment during a cold storage (8 °C, 90%–95% RH) for 5 days (simulating a short transport/storage) followed by a simulated commercialization period (at room temperature and ambient RH) up to 3 weeks. Three replicates (three packages) per each of 10 packaging treatments (5 packaging types×package activity (non-active and active)) were prepared per each of the following 4 sampling times: 5 days-cold storage (CS), CS+1 week (wk) of commercialization (CS+1 wk), CS+2 wk and CS+3 wk. Then, a total of 120 packages were prepared. All analyses and determinations were conducted in our laboratory of the Universidad Politécnica de Cartagena.

Figure 1. Package formats studied: (**A**) small tray (ST); (**B**) small box (SB); (**C**) large tray (LT); (**D**) large tray+alveoli (LT+); E, large box (LB).

2.3.2. Experiment 2: Industrial Validation

Once the effect of a short cold storage+commercialization period was studied in Experiment 1, the Experiment 2 studied the effect of the selected package type (LB, as described in "Results" section) during a long cold (8 °C) storage period up to 21 days (which may also simulate a transatlantic transport). For it, harvested mandarins were packaged in the LB (≈10 kg of fruit per cardboard box) (active or non-active) in the company (Fruca S.A.; Beniaján, Murcia, Spain) installations by the company personnel as usually conducted in pallets (83 cardboard boxes per pallet). Then, the packaged product was stored in the company cold rooms at 8 °C and 90%–95% RH.

Three replicates (three packages) per each 2 packaging treatments (non-active LB and active LB) were taken per each 3 sampling times: 7, 14, and 21 days. At each sampling time, three LB packages with fruit were taken from the company installations and transported to our laboratory of the Universidad Politécnica de Cartagena where all analyses and determinations (except weight loss, which was always monitored in the company cold rooms) were conducted.

2.4. Weight Loss

Weight of packages containing the product was monitored at each sampling time to determine the weight loss (%) of mandarins during storage.

2.5. Soluble Solids Content and Titratable Acidity

Juice from mandarin wedges was obtained with a blender (model MX2050; Braun, Germany). SSC was determined with a digital handheld refractometer (Atago N1; Tokyo, Kanto, Japan) at 20 °C and expressed as °Brix. TA of the diluted juice (2 mL plus 48 mL of distilled water) was determined with an automatic titrator (model T50; Metter Toledo; Milan, Italy) with 0.1 M NaOH to reach pH 8.1. *TA* was expressed as citric acid in %. *BrimA* index (Equation (1)) [5] was selected as the most appropriate maturity index for mandarins as recommended for citrus [6].

$$BrimA = SSC - (k \times TA) \tag{1}$$

where k is the tongue's sensitivity index, normally ranging from 2 to 10 [5], set at value 3 for citrus [6]. Each of the three replicates was analysed in duplicate.

2.6. Colour

Colour of mandarin (external colour) was determined using a colourimeter (Chroma Meter CR-400, Konica Minolta, Japan) at illuminant D65 and 2° observer, and with a viewing aperture of 8 mm. Three measurements were made (equatorial zone) per each fruit and they were then automatically averaged by the device. Ten mandarins were analysed per each replicate. The following specific colour index (Equation (2)) was selected as the most appropriate for orange citrus as previously reported [11,30]:

$$Colour\ index = \frac{1000 \times a^*}{L^* \times b^*} \tag{2}$$

where L^*, a^*, and b^* refer to the CIE colour parameters obtained from the colourimeter measurements.

2.7. Firmness

Firmness was determined with a texture analyzer (model TA XT Plus; TA Instruments; Surrey, UK) by measuring the amount of force (N) to compress 8-mm deep (P/10 probe of 10 mm Ø) a whole mandarin fruit on the diameter height. The texture analyzer was set for a drop speed of 20 mm min^{-1} and equipped with a load cell of 5 kg. Ten mandarins were analysed per each replicate.

2.8. Microbial Analyses and Decay Incidence

Surface microbial loads of mandarins were analysed as previously described [24]. Briefly, three mandarin fruits were mixed with buffered peptone water (1:1 w:volume) and then homogenised for 1 h at 120 rpm in an orbital shaker at 4 °C. Viable counts were based on duplicate counts by 10-fold serial dilutions in buffered peptone water. Then, aliquots (1 mL) of the microbial dilutions were pour-plated in plate count agar and violet red bile dextrose agar for mesophiles and psychrophiles and enterobacteria, respectively. For yeast and moulds, microbial aliquots (0.1 mL) were spread-plated on rose bengale agar. Mesophiles, psychrophiles, enterobacteria, yeast, and moulds were incubated at 31 °C (48 h), 4 °C (7 days), 37 °C (24 h), 25 °C (5 days) and 25 °C (7 days), respectively. Results were expressed as log colony forming units (CFU) cm^{-2}. Each of the three replicates was analysed in duplicate.

Decay incidence was quantified by calculating the percentage of rotten mandarins (fruits with visible mycelial growth) independently in each of replicates (packages).

2.9. Sensory Analyses

Sensory analyses were performed according to international standards [31]. Sensory tests were conducted in a standard room [32] equipped with ten individual taste booths. The panel consisted of twelve assessors (six women and six men, aged 22–61 years) who had been trained in discriminative quality attributes. Mandarin segments (5 units) were served at room temperature in transparent glass plates coded with three random digit numbers. Still mineral water was used as a palate cleanser. The quality attributes scored were overall quality (5: excellent; 3: fair, limit of acceptability (below 3 were not sensory accepted); 1: extremely bad), overall flavour (5: excellent; 3: fair; 1: extremely bad), sweetness (5: very sweet; 3: fair; 1: little sweet), acidity (5: very acid; 3: fair; 1: little acid) and juiciness (5: very juicy; 3: fair; 1: little juicy). The product shelf life was established based on the limit of acceptability of the product overall quality.

2.10. Statistical Analyses

The data were subjected to analysis of variance (ANOVA) using the SPSS software (v.19 IBM, New York, NY, USA). Statistical significance was assessed at $p = 0.05$, and the Tukey's multiple range test was used to separate the means.

3. Results

3.1. Experiment 1: Selection of Active Package (Pilot Plant Scale)

3.1.1. Weight Loss

The weight loss of samples after the cold storage period (CS; 5 days at 8 °C) was very low (<1.5%) (Table 1). No significant ($p > 0.05$) differences (SSC, TA, neither BrimA) among non-active and active samples were observed for any of the packaging treatments after the CS period (Table 1). Similarly, low weight losses have been reported in mandarins after short cold storage periods (<1 week) showing the adopted low storage temperature in cold rooms, together with high RH, a crucial role to slow down the physiological processes responsible for product weight losses (mainly due to dehydration) [33–35]. Nevertheless, mandarins are stored at room temperature for long periods during retail, leading to increased weight loss [36].

Table 1. Weight loss (%), soluble solid content (SSC; °Brix), titratable acidity (TA; %), BrimA maturity index (SSC-(3 × TA)), firmness (N) and colour index of mandarins packaged within different packages types (ST, small tray; SB, small box; LT, large tray; LT+, large tray with alveoli tray; and LB, large box), non-active (CT) and active, during a cold storage period (CS; 5 days at 8 °C) followed by a commercialization simulation (room temperature) up to 3 weeks (CS+1 wk, CS+2 wk and CS+3 wk) (n = 3 ± SD). Least significant differences are represented between parentheses.

	Package	Activity	Weight Loss	SSC	TA	BrimA	Firmness	Colour
	Day 0		-	15.1 ± 0.4	1.23 ± 0.12	11.4 ± 0.2	19.3 ± 1.5	7.8 ± 0.5
CS	ST	CT	0.3 ± 0.1	12.5 ± 0.5	1.58 ± 0.12	7.8 ± 0.2	16.5 ± 1.0	7.6 ± 1.0
		Active	0.3 ± 0.1	14.4 ± 1.4	1.48 ± 0.10	9.9 ± 1.3	17.6 ± 2.5	8.3 ± 0.3
	SB	CT	1.5 ± 0.5	13.4 ± 0.9	1.67 ± 0.13	8.4 ± 1.2	19.4 ± 2.0	7.7 ± 1.0
		Active	1.1 ± 0.2	12.9 ± 0.8	1.63 ± 0.03	8.0 ± 0.7	19.9 ± 2.8	7.9 ± 0.3
	LT	CT	0.8 ± 0.2	12.5 ± 0.5	1.62 ± 0.21	7.7 ± 0.7	20.4 ± 1.9	8.6 ± 0.6
		Active	0.7 ± 0.2	13.0 ± 0.8	1.52 ± 0.11	8.5 ± 0.6	20.0 ± 3.4	8.3 ± 0.8
	LT+	CT	0.9 ± 0.2	13.6 ± 0.5	1.59 ± 0.11	8.8 ± 0.6	20.7 ± 2.8	8.2 ± 0.6
		Active	1.3 ± 0.3	13.4 ± 0.7	1.48 ± 0.19	9.0 ± 1.1	18.8 ± 1.6	8.2 ± 0.7
	LB	CT	0.1 ± 0.0	13.9 ± 1.7	1.45 ± 0.09	9.5 ± 1.8	19.9 ± 3.3	8.1 ± 0.8
		Active	0.1 ± 0.0	12.7 ± 1.1	1.67 ± 0.10	7.7 ± 1.0	19.7 ± 2.1	8.3 ± 0.8
CS+1 wk	ST	CT	1.7 ± 0.2	13.2 ± 1.2	0.97 ± 0.12	10.3 ± 1.4	19.1 ± 1.9	6.9 ± 0.8
		Active	1.8 ± 0.3	13.0 ± 0.8	1.11 ± 0.15	9.7 ± 0.5	18.9 ± 1.0	7.2 ± 0.5
	SB	CT	5.7 ± 0.5	13.4 ± 1.3	1.10 ± 0.09	10.1 ± 1.3	18.4 ± 1.8	7.1 ± 0.9
		Active	4.5 ± 0.3	13.3 ± 1.1	1.11 ± 0.15	10.0 ± 1.1	18.4 ± 2.9	7.8 ± 0.7
	LT	CT	2.9 ± 0.4	13.5 ± 0.8	1.15 ± 0.08	10.0 ± 0.7	19.6 ± 3.1	6.8 ± 0.7
		Active	3.1 ± 0.4	13.0 ± 0.6	1.04 ± 0.15	9.9 ± 0.8	18.1 ± 1.9	6.8 ± 0.7
	LT+	CT	4.0 ± 0.5	13.4 ± 1.2	1.00 ± 0.18	10.4 ± 0.9	20.5 ± 2.8	7.2 ± 0.8
		Active	4.6 ± 0.5	13.2 ± 1.0	1.02 ± 0.10	10.1 ± 0.8	18.5 ± 1.5	7.4 ± 0.4
	LB	CT	2.60.4	14.3 ± 1.0	1.06 ± 0.13	11.1 ± 0.8	18.0 ± 2.5	7.5 ± 1.0
		Active	1.6 ± 0.1	14.6 ± 1.0	1.02 ± 0.15	11.5 ± 1.0	18.6 ± 1.5	7.3 ± 0.9
CS+2 wk	ST	CT	4.5 ± 0.9	11.5 ± 1.1	0.84 ± 0.09	9.0 ± 0.9	15.7 ± 1.7	8.0 ± 0.6
		Active	5.1 ± 0.7	14.4 ± 0.8	1.08 ± 0.15	11.2 ± 0.9	16.7 ± 1.3	8.1 ± 0.6
	SB	CT	11.4 ± 0.8	14.6 ± 0.4	0.85 ± 0.10	12.0 ± 0.5	14.2 ± 1.1	8.8 ± 0.7
		Active	10.2 ± 0.2	12.9 ± 1.1	1.06 ± 0.09	9.7 ± 1.3	15.0 ± 1.4	8.4 ± 0.4
	LT	CT	11.7 ± 1.2	13.5 ± 0.5	1.24 ± 0.13	9.8 ± 0.7	17.5 ± 1.5	8.6 ± 1.0
		Active	7.4 ± 1.1	14.1 ± 0.8	1.18 ± 0.07	10.5 ± 0.8	17.2 ± 1.6	8.3 ± 1.1
	LT+	CT	8.1 ± 0.3	13.3 ± 0.6	0.93 ± 0.08	10.5 ± 0.7	17.3 ± 1.0	8.2 ± 0.3
		Active	8.5 ± 0.4	13.8 ± 0.8	0.98 ± 0.08	10.9 ± 0.8	14.8 ± 1.2	8.1 ± 0.5
	LB	CT	8.7 ± 1.3	14.5 ± 1.1	1.15 ± 0.15	11.0 ± 1.1	16.4 ± 1.4	8.6 ± 0.9
		Active	5.9 ± 0.5	13.9 ± 1.0	1.22 ± 0.15	10.2 ± 0.8	17.1 ± 1.4	8.7 ± 1.0
CS+3 wk	ST	CT	11.0 ± 1.5	14.5 ± 0.5	0.93 ± 0.07	11.7 ± 0.6	16.7 ± 1.7	8.1 ± 0.5
		Active	8.2 ± 0.1	13.7 ± 1.0	0.96 ± 0.07	10.8 ± 0.9	17.6 ± 1.5	8.2 ± 0.3
	SB	CT	14.0 ± 0.9	16.0 ± 1.2	1.01 ± 0.07	13.0 ± 1.3	14.8 ± 1.0	8.7 ± 0.7
		Active	12.9 ± 0.3	13.9 ± 0.4	0.87 ± 0.05	11.4 ± 0.5	14.8 ± 1.4	8.8 ± 1.0
	LT	CT	15.9 ± 0.7	14.4 ± 0.4	0.90 ± 0.05	11.7 ± 0.4	17.5 ± 1.5	8.7 ± 0.8
		Active	9.9 ± 1.2	14.1 ± 0.4	0.90 ± 0.11	11.4 ± 0.2	14.8 ± 1.0	8.2 ± 0.7
	LT+	CT	10.2 ± 0.9	15.7 ± 0.4	0.93 ± 0.08	12.9 ± 0.3	15.4 ± 1.6	8.3 ± 0.7
		Active	10.9 ± 0.7	15.5 ± 0.4	1.04 ± 0.09	12.4 ± 1.2	15.0 ± 1.4	8.2 ± 1.0
	LB	CT	14.2 ± 1.2	15.2 ± 1.5	1.12 ± 0.11	11.8 ± 1.3	15.8 ± 1.5	8.6 ± 0.6
		Active	9.0 ± 0.5	13.5 ± 0.9	0.85 ± 0.09	10.9 ± 1.0	15.9 ± 1.3	8.7 ± 1.0
	Packaging type (A)		(0.7) ‡	(0.5) †	(0.05) *	(0.4) *	(0.7) †	(0.2) *
	Package activity (B)		(0.4) ‡	ns	ns	ns	ns	ns
	Storage time (C)		(0.6) ‡	(0.6) ‡	(0.08) ‡	(0.6) ‡	(0.9) ‡	(0.3) ‡
	A × B		(1.2) ‡	(0.8) ‡	ns	(0.8) ‡	(1.2) ‡	ns
	A × C		(1.4) ‡	(1.0) †	(0.18) ‡	(1.0) †	(1.9) ‡	(0.4) *
	B × C		(0.9) ‡	(0.6) †	(0.01) *	(0.5) *	ns	ns
	A × B × C		(2.0) ‡	(1.4) †	(0.25) ‡	(1.9) ‡	ns	ns

ns: not significant ($p > 0.05$); *, † and ‡ significance for $p \leq 0.05$, 0.01 and 0.001, respectively.

Weight loss of mandarins was increased during the commercialization period having all factors (packaging type, package activity and storage time), and their interactions, statistically significant effect on weight loss ($p < 0.001$) (Table 1). In accordance, mean (packaging type-package activity) weight losses of 3.3, 8.1, and 11.6 after CS+1 wk, CS+2wk, and CS+3wk, respectively, were observed. In particular, samples within non-active LB and LT showed the highest weight losses (14–16%) after CS+3wk, which may be attributed to the high product weight:package surface rates of these packages. Weight losses were significantly ($p < 0.001$) reduced using active packages, compared to non-active packages, with weight losses of 8%–9.5% after CS+3wk RT. Nevertheless, samples within active SB registered the highest weight loss (12.9%) among active samples after CS+3wk RT, probably due to the numerous and large perforations of this package that possibly intensified the weight loss in these mandarins. On the other side, the use of active LB and LT led to the highest weight loss reductions (decreases of 5.2 and 6.3 weight loss units, respectively) compared to their respective non-active samples after CS+3wk RT. In that sense, the controlled EOs release from active packages highly reduced the product weight loss for LB and LT. That finding denotes the benefits of the use of a macroperforated liner (LB and LT) compared to the macroperforated cardboard cover (SB), apart from the reduced visibility of the product for the consumer.

The transpirational water loss from the primary surface of plant products is limited by the plant cuticle as observed in citrus [37]. The chemical composition and/or the spatial arrangement of the components of the cuticle are more linked to other physical parameters like cuticle thickness or wax coverage [38]. Hence, the released EOs might change such cuticle properties, as previously studied [39], probably leading to the hereby observed lower weight loss of mandarins using the active packages.

3.1.2. Soluble Solids, and Titratable Acidity

Mandarins showed initial SSC and TA of 15.1 °Brix and 1.23%, respectively, at day 0 (Table 1). The individual factors package type and storage time were significant for SSC and TA, while package activity factor did not (Table 1). All factor interactions were also significant for SSC and TA (except packaging type×package activity for TA). SSC was decreased after the CS period by 1.6–2 °Brix showing LT and SB samples the highest SSC reductions with mean values (averaged between non-active and active) of 2.4 and 2 °Brix, respectively. In contrast, TA values of samples increased by 0.3–0.4 TA units after the CS period, without great differences (<0.2 TA units) between packaging treatments.

The cold storage of horticultural products is interpreted by plant cells as an abiotic stress leading to the synthesis of antioxidant compounds (e.g., organic acids), as previously reported in citrus fruit [40], during the first days of cold storage. Lately, the biosynthesis rates of these antioxidants decrease. In that sense, the observed TA increment and SSC reduction during cold storage may be explained by the organic acids biosynthesis using sugars as energy pools [15].

During the commercialization period, SSC increased, while TA was reduced, showing mandarins differences of 0.11–0.38 TA units after three weeks. Particularly, active LB samples showed the highest TA reduction of 0.38 TA units. Mature citrus fruits, which are classified as non-climacteric, evolve very low amounts of ethylene during ripening and a low respiration rate, but respond to exogenous ethylene by ripening-related pigment changes and accelerated respiration [41,42]. Furthermore, the ethylene biosynthesis pattern described as system I, typical from non-climateric fruit, revealed that the use of 1-Methylcyclopropene (a chemical competitive inhibitor of the ethylene active sites of plant products) in citrus resulted in an increase in ethylene evolution [41]. It has been demonstrated that EOs inhibited ethylene biosynthesis in several fruit and vegetables, although such mechanism is not still fully understood [43–48]. In contrast, it has been already demonstrated a competitive inhibition of EOs within the active sites of browning-relevant enzymes (polyphenol oxidase, peroxidase and phenylalanine ammonia-lyase) in lettuce [49]. In that sense, the inhibitory effect of EOs on ethylene production could be owed to a competitive inhibition of the EOs with the active sites of key enzymes of the ethylene biosynthesis pathway (aminocyclopropane-1-carboxylic acid synthase and Met-adenosyltransferase). Overall, EOs action could produce the same effect as 1-Methylcyclopropene in citrus [41] leading to an acceleration of senescence and respiration with the consequent consumption of organic acids and sugars increment (leading to an incremented maturity index as shown below) as reported in citrus fruits like mandarins [50,51]. TA differences >0.36 TA units have been reported to be significantly detected in sensory analyses [51]. Furthermore, reduced TA and increased SSC occurred during fruit postharvest life led to higher sensory scores and greater consumer acceptability of mandarins and oranges [6,51,52].

Due to the high importance of a balanced SSC-TA binomial for the fruit quality, the SSC/TA maturity index has been conventionally adopted. Nevertheless, BrimA index has been proposed as a better predictor of flavour for citrus fruits since SSC/TA is calculated as a ratio, rather than a subtractive calculation like BrimA [5,6]. The k value (Equation (1)) reflects the tongue's higher sensitivity to acid than to sugar allowing this index that smaller amounts of acid than sugar to make the same numerical change to BrimA, and in opposite direction. The BrimA index works on the simple principle that sugar and acid tend to have opposing effects on taste and the tongue perceives sugar and acids with differing sensitivities [5,53]. BrimA reached values of 11–13 at the third week of commercialization (Table 1). Particularly, active LB, LT and SB showed BrimA values below 11.5, showing active LB the lowest value (10.9), while the rest of the samples showed BrimA values above 11.5. Such trends are in accordance with sensory scores of active LB, LT, and SB after CS+3 wk period (see "Sensory analyses" section).

In conclusion, LB samples showed the highest TA reduction, which may be sensorially positive-appreciated, during commercialization while LT and SB samples showed the highest SSC reductions during the cold storage period, which may reduce their sensory acceptance.

3.1.3. Firmness

Mandarins showed an initial firmness of 19.3 N (Table 1). The individual factor storage time was significant ($p < 0.001$) for the fruit firmness. In that sense, a general decrease of ≈3 N was observed after the second week of commercialization when comparing mean values (average of all sample treatments) with day 0. The firmness reduction of mandarins, and other citrus fruits, during postharvest storage has been well correlated to the cellular wall polysaccharides, which may be decreased during fruit maturation [12]. The individual factor packaging type was also significant ($p < 0.01$) for firmness. Particularly, mean firmness values (averaged between active and non-active during all storage time) of LB samples showed the highest firmness (18.4 N), while SB showed the lowest mean firmness value (17.3 N). The benefit from using alveoli trays (LT+) on the fruit firmness was not hereby observed since such differences are only appreciated after vibrations occurred during a real transport. On the other side, the individual factor package activity was not significant ($p > 0.05$) for fruit firmness, neither the double interaction package type×package activity nor the triple interaction.

Overall, LB samples showed the highest firmness during storage, regardless of active or non-active packages. Nevertheless, our previous studies with tomatoes showed that product firmness was better maintained during storage when using active packages including the same EOs-βCD inclusion complex [24,25]. The latter finding may be explained since the tomato peel is more susceptible to dehydration processes during postharvest life than mandarin peel, which has the characteristic oil vesicles of citrus fruits that may provide higher mechanical resistance to the fruit.

3.1.4. Colour

Mandarins showed initial CIE colour parameters of $L^* = 59.5 \pm 1.0$, $a^* = 59.5 \pm 1.0$, $b^* = 59.5 \pm 1.0$, which corresponded to Chroma = 67.6 ± 2.7 and °Hue = 65.0 ± 1.6 (Supplementary Material Table S1). A marked luminosity (L^*) increase and a mild yellowness increment (b^*) were observed in the first week of commercialization after the cold storage period, which was correlated with a CI decrease after CS+1 wk (Table 1). Such colour changes are due to the accumulation of carotenoids (up to 94% present in the mandarin peel) after colour beak of the fruit, specially β-cryptoxanthin and apocarotenoids, which are responsible for the distinctive orange-reddish pigmentation in mandarins [54].

CI is a good colour index for citrus fruits that reflect all L^*, a^* and b^* changes together as previously reported [30]. It showed similar trends to a^*/b^* index, another classical colour index in citrus fruits [7]. The individual factor package type was significant ($p < 0.05$) for CI showing LT the highest colour changes after the first week of commercialization, while LB and LT+ showed the lowest CI changes. The observed CI differences among LT and LT+ may be explained since lower quantity of fruits, and more spaced, were disposed within the LT+ package probably leading to a lower ethylene accumulation, the ripening hormone with autocatalytic nature [55].

Later, as the commercialization period advanced, CI was increased to levels comparable to day 0 and CS. It may be explained by the mild redness (a^*) increment observed in such advanced commercialization stage, which agrees with the orange to red-orange turning during mandarin maturation, highly contributing a^* in the CI equation (Equation (2)). In contrast, package activity factor was not significant ($p > 0.05$) for colour parameters (L^*, a^*, b^* nor CI), neither the double, nor triple, interactions with this factor.

As observed, no great colour changes were observed during cold storage of mandarins, showing LB and LT+ the lowest colour changes after the first week of commercialization. Furthermore, the active material did not negatively affect the fruit colour during storage.

3.1.5. Sensory Analyses

Sensory quality of mandarins, and particularly flavour, is highly influenced by attributes like sweetness and acidity. Furthermore, the flavour acceptance of mandarins is the result of an appropriate high sweetness combined with a pleasant low-moderate acidity. In that sense, sensory analyses of this study were focussed on the determination of this complex attribute that is the fruit flavour. Meanwhile, other attributes like colour and consistency are more accurately determined by colour and firmness measurements, as previously showed.

After the CS period (Figure 2), all samples showed overall quality scores over the limit of acceptability, reaching the highest scores mandarins within active packages (4.2–4.5), while for non-active ones were closer to the limit of acceptability (3.4–3.9). Particularly, samples within active packages showed the highest sweetness scores (3.8–4.2) after CS period, without great differences among packaging treatments, showing active LB the lowest acidity score (2.7). The tongue reflects a higher sensitivity to acid than to sugar [53]. Furthermore, reduced acidity, together with increased sweetness, have been positively scored by panellists during maturation of oranges [6,51]. In that sense, the lower acidity in such active samples probably explains the higher sweetness appreciation of these samples after the CS period. Such reduced acidity of LB samples may be explained by the higher product weight:package surface rate of this package, which may lead to a higher autocatalytic synthesis of ethylene, with the consequent ripening effects, like the reduction of organic acids due to the product respiration that has been reported to occur in citrus fruit [50,51]. Nevertheless, ripening must be also controlled showing the active packaging a crucial role due to the reported properties of EOs to control ethylene biosynthesis in fruits [43]. No off-flavours, nor those related to EOs (due to the controlled and low EOs release from these active packages [20,24]), were detected during both CS and commercialization periods.

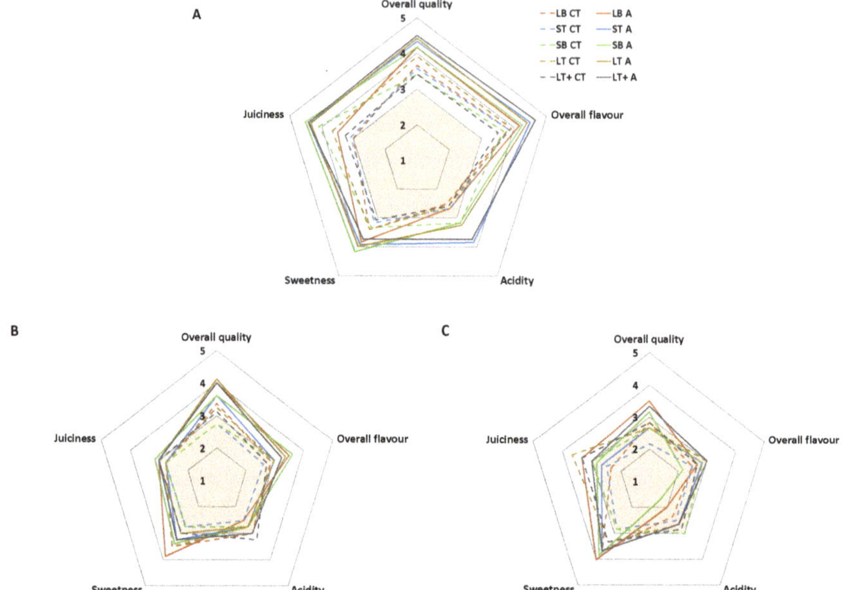

Figure 2. Sensory scores of mandarins packaged within different packages (ST, small tray; SB, small box; LT, large tray; LT+, large tray with alveoli tray; and LB, large box), non-active (CT) and active (A), after cold storage ((**A**); 5 days at 8 °C) followed by a commercialization simulation (room temperature) for 2 weeks (**B**) or 3 weeks (**C**). The coloured orange area defines the limit of acceptability (scores below 3 are considered below the limit of acceptability).

As expected, fruit maturation was enhanced during the commercialization period, characterized by a more pronounced acidity reduction coupled with a sweetness perception increment. Active samples showed higher sweetness and lower acidity compared to non-active samples. All non-active samples registered overall quality scores below the limit of acceptability after CS+3wk. Attending to active samples, those with higher product weight:package surface ratio (LB, LT and SB) reached overall quality scores over the limit of acceptability, displaying active LB the highest overall scores (3.5).

In conclusion, the shelf life of mandarins during a commercialization period (following a previous cold storage of five days at 8 °C) could be established in three weeks for active packages with high product weight:package surface ratio. Particularly, active LB reached the highest sensory scores, while it is reduced to two weeks for non-active packages.

3.1.6. Microbiology and Decay Incidence

Mandarins showed initial mesophilic, psychrophilic, enterobacteria, yeasts and moulds loads of 1.3, 1, <0.5, 2.3 and 2.2 log CFU cm^{-2}, respectively (Table 2). In general, microbial loads increased during storage periods, although such microbial growth was low. In that sense, microbial loads of samples were below 2.8 (mesophilic), 2.6 (psychrophilic), 1.6 (enterobacteria), 2.4 (yeasts) and 3.5 log CFU cm^{-2} (moulds) after CS+3wk. Such low surface microbial loads are usual in fresh fruits, compared to processed (e.g., fresh-cut) fruits. The higher microbial loads corresponded to moulds. In particular, *Alternaria* and *Cladosporium* have been reported as the main moulds in mandarins, followed by *Fusarium* and *Rhizopus* [56].

Table 2. Microbial loads (log CFU cm^{-2}) of mandarins packaged within different packages types (ST, small tray; SB, small box; LT, large tray; LT+, large tray with alveoli tray; and LB, large box), non-active (CT) and active, during a cold storage period (CS; 5 days at 8 °C) followed by a commercialization simulation (room temperature) up to 3 weeks (CS+1 wk, CS+2 wk and CS+3 wk) ($n = 3 \pm$ SD). Least significant differences are represented between parentheses.

Storage	Package	Activity	Mesophiles	Psychrophiles	Enterobacteria	Yeast	Moulds
	Day 0		1.3 ± 0.1	1.0 ± 0.1	<0.5	2.3 ± 0.5	2.2 ± 0.3
CS	ST	CT	1.9 ± 0.1	1.0 ± 0.7	0.5 ± 0.2	2.3 ± 0.4	1.9 ± 0.4
		Active	1.6 ± 0.2	0.6 ± 0.8	0.5 ± 0.3	2.3 ± 0.1	1.5 ± 0.3
	SB	CT	2.3 ± 0.1	1.9 ± 0.2	1.2 ± 0.1	2.0 ± 0.2	2.3 ± 0.1
		Active	1.8 ± 0.2	1.1 ± 0.2	<0.5	1.9 ± 0.3	1.8 ± 0.1
	LT	CT	2.5 ± 0.3	2.5 ± 0.3	1.6 ± 0.2	2.4 ± 0.1	2.3 ± 0.2
		Active	1.9 ± 0.2	0.8 ± 0.2	<0.5	1.9 ± 0.3	2.1 ± 0.2
	LT+	CT	2.3 ± 0.4	1.1 ± 0.4	0.5 ± 0.2	2.0 ± 0.2	2.0 ± 0.1
		Active	1.7 ± 0.3	0.5 ± 0.4	<0.5	1.5 ± 0.5	2.1 ± 0.1
	LB	CT	1.9 ± 0.4	2.1 ± 0.8	<0.5	2.2 ± 0.1	1.8 ± 0.2
		Active	1.2 ± 0.2	1.4 ± 0.2	<0.5	1.7 ± 0.4	1.7 ± 0.2
CS+1 wk	ST	CT	2.4 ± 0.3	2.0 ± 0.4	0.8 ± 0.1	2.4 ± 0.2	2.7 ± 0.1
		Active	2.0 ± 0.4	1.6 ± 0.2	0.5 ± 0.4	1.7 ± 0.3	2.4 ± 0.1
	SB	CT	2.6 ± 0.2	2.0 ± 0.6	1.5 ± 0.1	1.9 ± 0.3	2.7 ± 0.2
		Active	2.1 ± 0.4	1.3 ± 0.2	<0.5	1.7 ± 0.4	2.4 ± 0.2
	LT	CT	2.5 ± 0.2	2.3 ± 0.5	1.7 ± 0.1	2.0 ± 0.3	2.6 ± 0.2
		Active	1.7 ± 0.3	1.7 ± 0.2	0.5 ± 0.1	2.0 ± 0.1	2.6 ± 0.1
	LT+	CT	2.4 ± 0.2	1.2 ± 0.1	1.1 ± 0.4	2.3 ± 0.1	1.9 ± 0.2
		Active	2.0 ± 0.2	0.2 ± 0.1	<0.5	1.7 ± 0.1	1.8 ± 0.2
	LB	CT	2.3 ± 0.1	1.6 ± 0.6	0.8 ± 0.2	2.3 ± 0.1	2.5 ± 0.3
		Active	1.3 ± 0.3	0.6 ± 0.3	<0.5	2.0 ± 0.2	2.0 ± 0.2
CS+2 wk	ST	CT	2.5 ± 0.1	1.2 ± 0.1	1.3 ± 0.3	2.2 ± 0.1	3.0 ± 0.3
		Active	1.9 ± 0.1	1.1 ± 0.4	<0.5	1.9 ± 0.2	3.0 ± 0.2
	SB	CT	2.7 ± 0.3	2.1 ± 0.2	1.2 ± 0.2	2.0 ± 0.2	2.6 ± 0.2
		Active	1.9 ± 0.3	1.4 ± 0.2	0.5 ± 0.2	1.6 ± 0.4	2.2 ± 0.1
	LT	CT	2.3 ± 0.3	1.4 ± 0.2	1.2 ± 0.3	1.5 ± 0.5	2.5 ± 0.1
		Active	2.1 ± 0.3	1.5 ± 0.9	<0.5	1.5 ± 0.3	2.1 ± 0.2
	LT+	CT	2.1 ± 0.2	2.1 ± 0.5	1.2 ± 0.2	2.0 ± 0.2	2.7 ± 0.3
		Active	1.3 ± 0.4	1.4 ± 0.2	<0.5	1.3 ± 0.2	2.0 ± 0.3
	LB	CT	2.6 ± 0.3	1.6 ± 0.2	0.5 ± 0.4	2.3 ± 0.2	2.6 ± 0.1
		Active	1.6 ± 0.1	1.8 ± 0.2	0.6 ± 0.4	1.2 ± 0.1	2.2 ± 0.1
CS+3 wk	ST	CT	2.8 ± 0.3	2.2 ± 0.4	1.1 ± 0.2	2.2 ± 0.2	3.1 ± 0.2
		Active	1.9 ± 0.2	2.1 ± 0.3	0.9 ± 0.2	2.1 ± 0.1	2.3 ± 0.2
	SB	CT	2.3 ± 0.1	2.5 ± 0.2	1.6 ± 0.4	1.7 ± 0.2	2.0 ± 0.1
		Active	1.8 ± 0.2	1.5 ± 0.3	<0.5	1.2 ± 0.1	1.9 ± 0.1
	LT	CT	2.6 ± 0.3	2.6 ± 0.5	0.9 ± 0.2	2.0 ± 0.4	3.4 ± 0.2
		Active	1.8 ± 0.2	1.7 ± 0.1	<0.5	1.3 ± 0.2	2.4 ± 0.3
	LT+	CT	2.3 ± 0.2	1.9 ± 0.3	1.0 ± 0.1	1.8 ± 0.2	2.5 ± 0.2
		Active	1.7 ± 0.2	1.0 ± 0.1	<0.5	1.2 ± 0.1	1.9 ± 0.3
	LB	CT	2.6 ± 0.2	1.8 ± 0.4	<0.5	2.4 ± 0.2	3.5 ± 0.2
		Active	1.9 ± 0.2	1.3 ± 0.2	0.5 ± 0.3	2.0 ± 0.4	3.0 ± 0.3
Packaging type (A)			(0.2) ‡	(0.4) ‡	(0.2) ‡	(0.3) ‡	(0.2) ‡
Package activity (B)			(0.1) ‡	(0.2) ‡	(0.1) ‡	(0.2) ‡	(0.1) ‡
Storage time (C)			(0.2) ‡	(0.4) ‡	(0.2) ‡	(0.3) ‡	(0.2) ‡
A×B			ns	ns	(0.2) ‡	ns	ns
A×C			(0.5) ‡	(0.8) ‡	(0.4) ‡	(0.6) ‡	(0.4) ‡
B×C			(0.3) ‡	(0.5) ‡	(0.2) ‡	(0.4) ‡	(0.3) ‡
A×B×C			ns	ns	(0.5) ‡	ns	(0.6) ‡

ns: not significant ($p > 0.05$); ‡, significance for $p \leq 0.001$.

All individual factors (package type, package activity and storage time) were significant ($p < 0.001$) for all microbial groups. In general, samples within active packages showed lower microbial loads than their respective non-active packaged samples. Furthermore, double interactions with storage time were also significant ($p < 0.001$) for all microbial groups.

Mesophilic loads augmented by 0.5–1 log units after the CS period, showing active LB the lowest load with 1.2 log CFU cm^{-2}. Mesophilic loads of non-active packaged samples incremented by 1–1.1, 1.2–1.4 and 1.4–1.6 log units after one, two, and three weeks of commercialization, respectively, for LB and ST packages. Meanwhile, the rest of non-active packaged mandarins registered increments of 1–1.4 during the three-week commercialization without high differences among them. Although non-active LB and non-active ST samples showed the highest mesophilic increments, their respective active packages highly reduced the mesophilic increments after CS+3 wk by two and 2.7 fold, respectively. In that sense, samples packaged within active packages showed similar ($p > 0.05$) mesophilic loads of 1.7–1.9 log CFU cm^{-2} after CS+3 wk.

In general, no significant ($p > 0.05$) psychrophilic increments were observed after the CS period, except non-active LT and non-active LB that registered psychrophilic growth of 1.4–1.5 log units. The higher growth in these packaging treatments may be explained by the high quantity of samples per package, which may stimulate the microbial growth due to the high humidity and low ventilation, especially in the fruits in the bottom of the package. Nevertheless, the active LT and active LB packages showed 1–1.7 lower log units after the CS period, compared to their corresponding non-active packages. No significant ($p > 0.05$) psychrophilic growth was observed during the first 1 week of commercialization. At the third week of commercialization, psychrophilic increments of 1–1.5 were observed for non-active ST, SB and LT, compared to their respective initial loads, although their respective active versions did not register significant changes ($p > 0.05$). Attending to their final loads, active LB and active LT+ registered the lowest psychrophilic loads of 1–1.3 log CFU cm^{-2} after CS-3wk.

Enterobacteria loads were below 1 log CFU cm^{-2} after the CS period, except non-active SB and LT with loads of 1.2 and 1.6 log units, respectively. After the first week of commercialization, enterobacteria loads of non-active packages increased by 1–1.7 log units, while no significant (p > 0.05) changes were observed for active packages. Enterobacteria loads did not change (p > 0.05) during the second and third week of commercialization, compared to the first week, being kept the latter trend related to the effectiveness of active packages. In that sense, active packages showed enterobacteria loads below 1 log CFU cm^{-2} after 3 weeks of commercialization.

Mould loads of samples did not register high changes (<0.8 log units) after CS, CS+1 wk and CS+2 wk periods. Non-active LB, ST, and LT showed mould increments of 1–1.3 log CFU cm^{-2} after CS+3 wk, compared to their respective initial levels. In general, active samples showed 0.5–1 fewer log units than their respective non-active samples at CS+3 wk. Attending to yeasts, low changes (<1 log units) were observed during both cold storage and three weeks of commercialization. As observed, low yeast and moulds changes were observed during storage periods of mandarins due to the lower growth rate of these microbial groups. Attending to decay incidence, the storage time factor was not significant ($p > 0.05$), neither their double and triple interactions with package type and package activity (data not shown). Mean values of samples during the complete commercialization period did not show significant ($p > 0.05$) differences among them, with decay incidence values ranging between 1.3% and 7%. The absence of the expected increment of decay incidence throughout commercialization may be due to the low quantity of fruits per packages, especially for ST, SB and LT+. For the same reason, no conclusions can be obtained related to the effect of package type and package activity factors on the decay incidence of samples.

As observed, no microbial growth (except mesophilic loads that slightly increased by 0.5–1 units) was observed after the 5-days cold storage at 8 °C, which remarks the importance of cold storage to extend the shelf life of citrus fruit attending to microbial quality [13]. Nevertheless, fruits are usually commercialized during retail at temperatures higher than 8 °C, usually at room temperature for the case of citrus fruits. Particularly, microbial growth of 1–1.6 log units was achieved after 3 weeks of commercialization simulation (with the previous cold storage) reaching mesophiles the highest increments at this room temperature period. Nevertheless, active packages highly controlled the microbial growth, especially high in non-active LB, but active LB better controlled the microbial growth among the rest of active packages. In that sense, the controlled EO release from the βCD inclusion complex (as characterized in our previous publications [20,24]) led to an antimicrobial effect, which was more appreciable during the commercialization period since at such higher temperatures the EO release from the inclusion complex is higher. The latter behaviour is explained since molecular Brownian motion is enhanced with the temperature increments, leading to a higher EO release from the inclusion complex [57].

As far as microbial growth is concerned, enterobacteria seemed to be more sensitive to the released EOs from active packages, as previously observed in tomatoes by our Group [20,28]. Carvacrol, the major component of the hereby used EOs mix of the complex, was considered, together with thymol (the major component of oregano EO together with carvacrol), as the EOs component with the widest spectra activity. In that sense, carvacrol is effective against both gram-negative and gram-positive with even higher effectiveness against the gram-negative bacteria (e.g., enterobacteria) [17,58]. In that sense, the hydroxyl group from the structure of this phenolic compound is crucial to disintegrate the outer membrane of gram-negative bacteria, which is more susceptible to the EOs antimicrobial properties [18]. Furthermore, the higher susceptibility of gram-negative bacteria to EOs has been hypothesized due to the less dense cell wall and lower peptidoglycan content compared to Gram-positive bacteria, as observed in a study with oregano and thyme EOs [58].

3.2. Experiment 2: Industrial Validation of the Selected Active Packaging

In accordance with experiment 1, LB was selected for the industrial validation experiment owed to the high beneficial effects of active LB on the mandarin quality to reach a long shelf life. Once the quality changes were studied during the most adverse storage conditions for mandarins (i.e., room temperature storage), the industrial validation experiment was studied at a recommended cold storage for mandarins [13].

3.2.1. Weight Loss

The weight loss of samples after 7 days was very low (<1%), as previously observed in the experiment 1. Package activity and storage time individual factors were significant ($p < 0.001$) for the weight loss of samples (Figure 3). In that sense, samples within active packages showed 1–1.1 fewer weight loss units than non-active packaged mandarins during storage. As expected, weight losses increased during storage with an increment of 4.2–4.3 weight loss units after 14 and 21 days, respectively. Nevertheless, no significant differences ($p > 0.05$) were observed between non-active and active samples.

Overall, weight losses of samples within non-active and active packages during storage were low after 14 days (2.7% and 1.6%, respectively). Weight losses were then incremented to 6.9% and 5.9% for non-active and active samples, respectively, after 21 days at 8 °C. As previously discussed, the released EOs may cause mild structural changes in the structures of the fruit surface [39], such as in the cuticle, leading to a lower water mass transport in accordance with weight loss data.

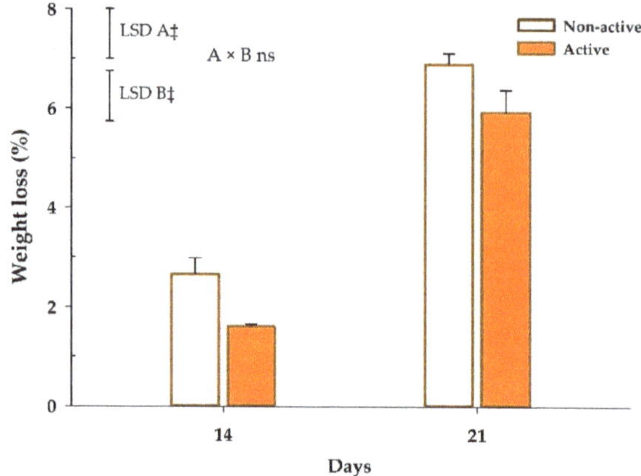

Figure 3. Weight loss of mandarins packaged within non-active and active packages up to 21 days at 8 °C (*n* = 3, ± SD). Least significant differences (LSD) are represented with bars. The uppercase letters A and B denote package type and storage time, respectively. ns, not significant (*p* > 0.05). ‡ significance for *p* ≤ 0.001.

3.2.2. Soluble Solids Content, Titratable Acidity and Firmness

Mandarins showed initial SSC, TA, and BrimA values of 14.3 °Brix, 0.9%, and 11.6, respectively, at day 0 (Figure 4A–C). Package activity and storage time effects, as well as their interaction, were not significant (*p* > 0.05) for SSC, with SSC values of 13.7–15.1 °Brix during all cold storage period (Figure 4A). The observed slight SSC reduction during the five day-cold storage period (experiment 1) was not hereby observed, since this antioxidant response, which implies the use of sugars as an energy source [15], was probably downregulated from day 5 to day 7.

As far as TA is concerned, the package activity, and its interaction with storage time, did not show significant (*p* > 0.05) effects on TA (Figure 4B). In general, no significant changes (*p* > 0.05) were observed between active and non-active samples although non-active samples showed a TA value 0.18 units lower after 21 days of storage. Samples packaged within the active package did not show (*p* > 0.05) such TA decrease, but a similar trend was observed. This TA decrease is common during citrus fruit maturation, as previously discussed due to the respiratory activity of the plant product [50,51], and it was positively correlated with a higher sensory acceptability of mandarins [6,51,52]. Such respiratory rates are higher at room temperature compared to the cold storage temperature, since a significant TA reduction was only observed after 21 days at 8 °C.

As expected, the ripening of mandarins advanced during cold storage in accordance with BrimA values (*p* < 0.01), although in a lower rate compared to the commercialization period (experiment 1). In that sense, significant BrimA changes were only observed after 21 days of storage at 8 °C (Figure 4C). Contrary to storage time, the package activity factor did not show a significant effect (*p* > 0.05) on BrimA maturity index, neither the package activity×storage time interaction.

Mandarins showed firmness values of 16.2–18.7 N during cold storage, with no significance (*p* > 0.05) for any of the individual factors and their interaction (data not shown).

Conclusively, the use of the active package did not affect the SSC and TA attributes during storage at 8 °C, which may not compromise the consumer acceptance due to the high importance of these attributes on the flavour acceptance of mandarin samples [6]. Furthermore, the firmness of mandarins was not affected during cold storage, independently of the packaging treatment.

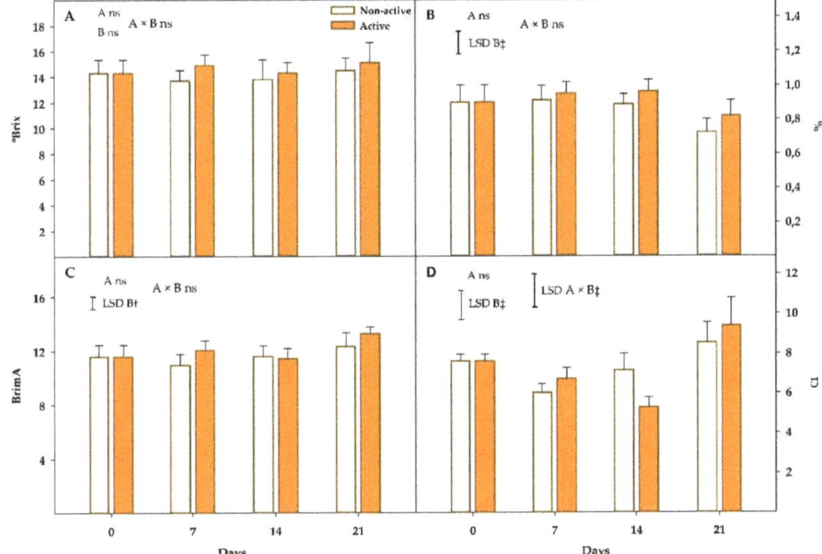

Figure 4. Soluble solids content (**A**), titratable acidity (**B**), BrimA maturity index (**C**) and colour index (**D**) of mandarins packaged within non-active and active packages up to 21 days at 8 °C (n = 3 ± SD). Least significant differences (LSD) are represented with bars. The uppercase letters A and B denote package type and storage time, respectively. ns, not significant ($p > 0.05$). † and ‡ significance for $p \leq 0.01$ and 0.001, respectively.

3.2.3. Colour

Mandarins showed initial CIE colour parameters of $L^* = 64.1 \pm 0.7$, $a^* = 30.0 \pm 1.3$, $b^* = 61.6 \pm 1.3$ in this experiment 2, which corresponded to Chroma = 68.5 ± 1.5 and °Hue = 64.0 ± 0.9 (Supplementary material Table S2). The storage time factor was significant ($p < 0.001$) for all L^*, a^* and b^*. These colour parameters were reduced during storage. Particularly, L^* and a^* showed low and moderate reductions of <3 and <9 units after 21 days of cold storage, respectively, while b^* was reduced by 6 and ≈20 units after 7–14 and 21 days, respectively. These colour changes corresponded to an increment of CI during cold storage, reaching CI increases in active samples of 1.8–2 units after 14–21 days (Figure 4D), while such CI increments were not significant ($p > 0.05$) for non-active samples. As observed, active packages still allowed a moderate fruit turning from orange to red-orange, as a result of the carotenoids biosynthesis [54]. Nevertheless, these colour changes were not high, which would not compromise the consumer acceptance (see "Sensory analyses" section).

3.2.4. Sensory Analyses

Sensory analyses of samples after 14 days of storage showed good sensory scores without high differences (<0.5) between samples stored within non-active and active packages (Figure 5). Particularly, mean values (non-active and active) for overall quality, overall flavour, acidity, sweetness and juiciness of 4.4, 3.6, 3, 4, and 3.9, respectively, were observed after 14 days.

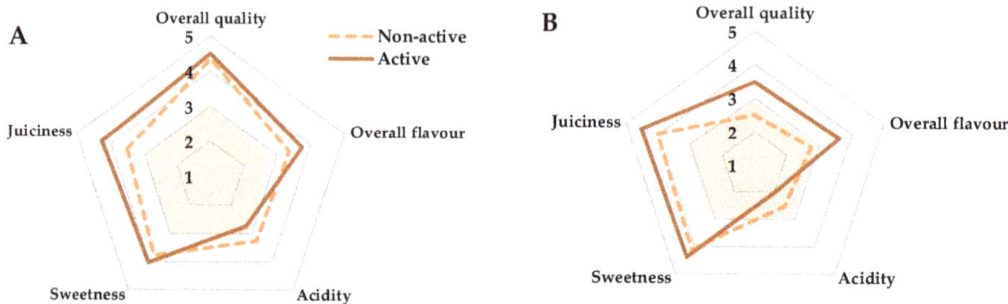

Figure 5. Sensory scores of mandarins packaged within active or non-active package after 14 days (**A**) and 21 days (**B**) of storage at 8 °C. The coloured orange area defines the limit of acceptability (scores below 3 are considered below the limit of acceptability).

After 21 days, samples within active packages still showed sensory scores over the limit of acceptability, with overall quality and overall flavour scores of 3.5 and 3.6, respectively, while non-active samples were not sensorially accepted (Figure 5). Furthermore, active samples showed lower acidity scores than non-active samples, which is in accord with TA data, which might lead to a higher consumer acceptance [6,51].

In conclusion, the shelf life of mandarins during storage at 8 °C could be established, based on sensory quality related to flavour attributes, in at least 21 days for samples stored within the active package, while such shelf life is reduced to 14 days when mandarins are stored within non-active packages.

3.2.5. Microbiology and Decay Incidence

Mandarins also showed low initial loads (1–1.5 log CFU m^{-2}) for all microbial groups (Figure 6), as previously explained in Experiment 1. The storage time factor was significant ($p < 0.001$) for all microbial groups, except for psychrophiles.

Attending to mesophiles, an increment of 1 log unit was observed after seven days in samples stored within the non-active package, whose loads were maintained ($p > 0.05$) during the rest of storage period (Figure 6A). Package activity factor was significant ($p < 0.01$) for mesophiles, with loads ≈1.7 log units lower at every sampling time compared to the non-active samples.

Psychrophiles loads of 0.5–1.3 log CFU m^{-2} were observed during all storage periods without significant ($p > 0.05$) changes among the different sampling times (data not shown). Furthermore, package activity factor was not significant ($p > 0.05$) for psychrophiles loads.

Enterobacteria growth was also very low during storage with the only significant increment observed at day 7 (Figure 6B). Package activity factor was significant ($p < 0.01$) for enterobacteria loads, showing samples within active packages 0.5–1.1 lower log units than non-active samples. As observed, enterobacteria growth at 8 °C was low due to the higher optimum growth temperatures for this microbial group. Although such lower enterobacteria growth at this storage temperature did not allow to show the higher effectivity of EOs against this microbial group, as observed in Experiment 1, mandarins within active packages showed the lowest enterobacteria loads (0.6 log CFU m^{-2}) after 21 days at 8 °C.

Moulds showed the highest growth among microbial groups with 2 and 1.5 higher log units after 21 days for non-active and active samples, respectively (Figure 6C). Particularly, the highest growth was observed from day 7 to day 21, while no significant growth ($p > 0.05$) was detected in the first seven days of storage. As observed, the package activity factor was significant for mould loads, showing mandarins within the active package ≈0.6 lower log units than non-active samples at days 14 and 21. The lower optimum growth temperature for moulds allowed the observed higher growth, compared to other groups like enterobacteria, although active packages still controlled such mould growth.

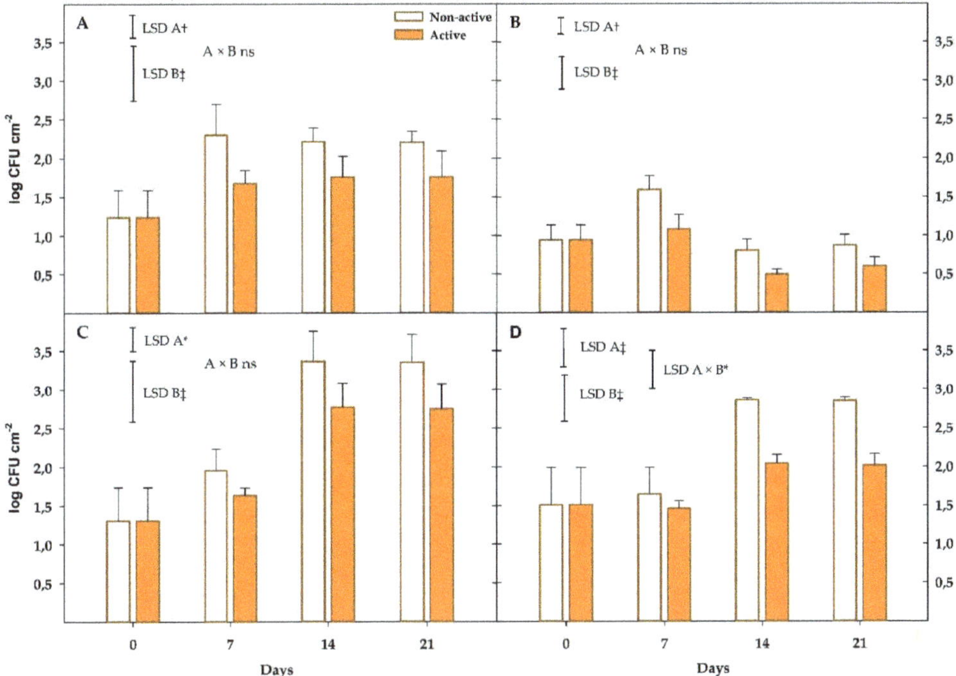

Figure 6. Microbial loads (**A**), mesophiles; (**B**), enterobacteria; (**C**), moulds; (**D**), yeasts of mandarins packaged within non-active and active package up to 21 days at 8 °C ($n = 3$, ± SD). Least significant differences (LSD) are represented with bars. The uppercase letters A and B denote package type and storage time, respectively. ns, not significant ($p > 0.05$). *, † and ‡ significance for $p \leq 0.05$, 0.01 and 0.001, respectively.

Similar to moulds, package activity and storage time factors were significant ($p < 0.001$) for yeasts (Figure 6D). In that sense, yeast growth of 1.4 log units was observed in non-active samples, while increments of only 0.5 units were observed in active samples after 21 days.

Decay incidences of 1.1%–1.6%, 5.5%–6.5% and 10.4%–17.4% were observed after seven, 14 and 21 days, respectively (Figure 7). Package activity and storage time factors, as well as their interaction, were significant for decay incidence ($p < 0.001$). Mandarins within active packages showed lower decay incidence that non-active samples. Furthermore, this benefit was even enhanced as the storage time increased with a two-fold reduction of decay incidence after 21 days.

In conclusion, the controlled release of EOs from the βCD inclusion complex of active boxes allowed to control microbial growth of mandarins packaged within this active package, which was also reflected on the reduced decay incidence. Furthermore, the inhibition of ethylene production by EOs [43,44] may also be responsible for the observed decay incidence reduction since ethylene may promote fungus growth as widely reported [59].

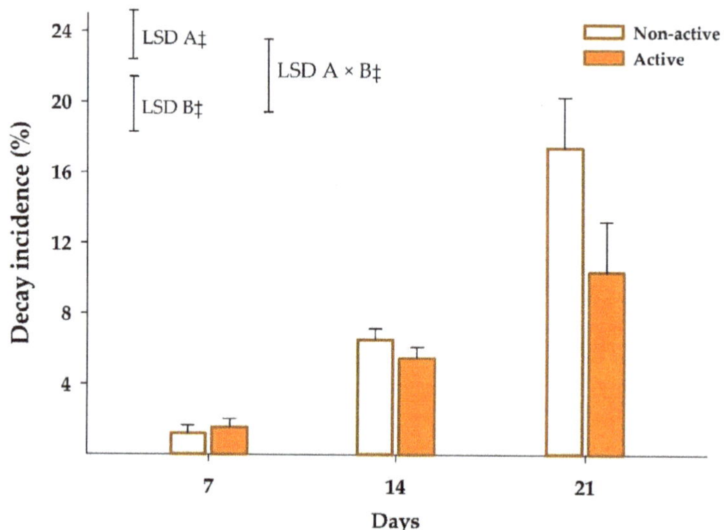

Figure 7. Decay incidence of mandarins packaged within non-active and active package up to 21 days at 8 °C (*n* = 3, ± SD). Least significant differences (LSD) are represented with bars. The uppercase letters A and B denote package type and storage time, respectively. ‡ significance for $p \leq 0.001$.

4. Conclusions

The use of active packaging including βCD-EOs inclusion complex extended the shelf life of mandarins during a commercialization simulation (room temperature; conducted after a short storage period simulating a short transport/storage) from two weeks (non-active packages) to three weeks. Different package formats were studied (different sized trays and boxes, and even including alveoli trays), with "large box" format ($\approx$10 kg fruit per box) showing the best results during this commercialization period. Particularly, the active large box better controlled the microbial growth of mandarins (with loads up to two log units lower after three weeks) and colour changes, while it induced the highest reductions of weight losses and the greatest acidity decrease of mandarins after three weeks (correspondent to the lowest maturity index changes among active samples), which was highly appreciated in the sensory analyses. This package format was then selected and validated at the industrial level during a long storage (up to 21 days) of mandarins. The active large box showed the same benefits, reducing microbial growth (together with decay incidence) and weight losses while maintaining the physicochemical quality (soluble solids, titratable acidity, firmness, and colour). In that sense, the shelf life of mandarins was extended from 14 days (non-active large box) to at least 21 days. As observed, the controlled release of EOs from the active large box extended the shelf life of mandarins either during a long cold transportation simulation or a commercialization period at non-recommended room temperature.

Supplementary Materials: The following supplementary materials are available online at http://www.mdpi.com/2304-8158/9/5/590/s1. Table S1: CIE colour parameters (*L*, *a*, *b*; Chroma, °Hue) of mandarins (Experiment 1), Table S2: CIE colour parameters (*L*, *a*, *b*; Chroma, °Hue) of mandarins (Experiment 2).

Author Contributions: Conceptualization, A.L.-G.; methodology, A.L.-G., L.B.-M., L.N.-S. and M.R.-C.; validation, A.L.-G.; formal analysis, G.B.M.-H.; investigation, all authors; resources, L.B.-M. and A.L.-G.; data curation, M.R.-C., L.B.-M. and G.B.M.-H.; writing—original draft preparation, G.B.M.-H.; writing—review and editing, G.B.M.-H. and A.L.-G.; visualization, G.B.M.-H.; supervision, A.L.-G.; project administration, A.L.-G. and M.R.-C.; funding acquisition, A.L.-G. All authors have read and agreed to the published version of the manuscript.

Funding: This research was funded through a European Union's Horizon 2020 research and innovation programme under Grant Agreement No 812001.

Acknowledgments: To the companies Blancasol and Fruca for supplying the mandarin fruits used in the experiments and the use of their installations.

Conflicts of Interest: The authors declare no conflict of interest.

References

1. FAOSTAT. Available online: http://www.fao.org/faostat/en/#data (accessed on 17 October 2018).
2. Palou, L.; Valencia-Chamorro, S.; Pérez-Gago, M. Antifungal edible coatings for fresh citrus fruit: A review. *Coatings* **2015**, *5*, 962–986. [CrossRef]
3. Lee, S.K.; Kader, A.A. Preharvest and postharvest factors influencing vitamin C content of horticultural crops. *Postharvest Biol. Technol.* **2000**, *20*, 207–220. [CrossRef]
4. Campbell, B.L.; Nelson, R.G.; Ebel, R.C.; Dozier, W.A. Mandarin attributes preferred by consumers in grocery stores. *HortScience* **2006**, *41*, 664–670. [CrossRef]
5. Jordan, R.B.; Seelye, R.J.; McGlone, V.A. A sensory-based alternative to brix/acid ratio. *Food Technol.* **2001**, *55*, 36–44.
6. Obenland, D.; Collin, S.; Mackey, B.; Sievert, J.; Fjeld, K.; Arpaia, M.L. Determinants of flavor acceptability during the maturation of navel oranges. *Postharvest Biol. Technol.* **2009**, *52*, 156–163. [CrossRef]
7. Rodrigo, M.J.; Marcos, J.F.; Zacarías, L. Biochemical and molecular analysis of carotenoid biosynthesis in flavedo of orange (*Citrus sinensis* L.) during fruit development and maturation. *J. Agric. Food Chem.* **2004**, *52*, 6724–6731. [CrossRef]
8. Campbell, B.L.; Nelson, R.G.; Ebel, R.C.; Dozier, W.A.; Adrian, J.L.; Hockema, B.R. Fruit quality characteristics that affect consumer preferences for satsuma mandarins. *HortScience* **2004**, *39*, 1664–1669. [CrossRef]
9. Stewart, I.; Wheaton, T.A. *Effects of Ethylene and Temperature on Carotenoid Pigmentation of Citrus Peel*; Food and Agriculture Organization: Rome, Italy, 1971.
10. Harding, P.L.; Sunday, M.B. Seasonal changes in Florida temple oranges. *Usda Tech. Bull.* **1953**, *1072*, 1–61.
11. Jiménez-Cuesta, M.; Cuquerella, J.; Martínez-Jávega, J.M. Determination of a color index for citrus fruit degreening. In Proceedings of the International Society of Citriculture, Tokyo, Japan, 9–12 November 1981; Volume 2, pp. 750–753.
12. Muramatsu, N.; Takahara, T.; Kojima, K. Relationship between texture and cell wall polysaccharides of fruit flesh in various species of citrus. *HortScience* **1996**, *31*, 114–116. [CrossRef]
13. Kader, A.A.; Arpaia, M.L. Postharvest handling systems: Subtropical fruits. In *Postharvest Technology of Horticultural Crops*; Kader, A.A., Ed.; University of California, Division of Agriculture and Natural Resources: San Diego, CA, USA, 2002; pp. 375–384.
14. Dorman, H.J.D.; Deans, S.G. Antimicrobial agents from plants: Antibacterial activity of plant volatile oils. *J. Appl. Microbiol.* **2000**, *88*, 308–316. [CrossRef]
15. Buendía-Moreno, L.; Soto-Jover, S.; Ros-Chumillas, M.; Antolinos-López, V.; Navarro-Segura, L.; Sánchez-Martínez, M.J.; Martínez-Hernández, G.B.; López-Gómez, A. An innovative active cardboard box for bulk packaging of fresh bell pepper. *Postharvest Biol. Technol.* **2020**, *164*, 111171. [CrossRef]
16. López-Gómez, A.; Boluda-Aguilar, M.; Soto-Jover, S. The use of refrigerated storage, pretreatment with vapors of essential oils, and active flow-packing, improves the shelf life and safety of fresh dill. In Proceedings of the 24th IIR International Congress of Refrigeration, Yokohama, Japan, 16–22 August 2015; pp. 4554–4560.
17. Zaika, L.L. Spices and herbs: Their antimicrobial activity and its determination. *J. Food Saf.* **2007**, *9*, 97–118. [CrossRef]
18. Burt, S. Essential oils: Their antibacterial properties and potential applications in foods—A review. *Int. J. Food Microbiol.* **2004**, *94*, 223–253. [CrossRef] [PubMed]
19. Rao, J.; Chen, B.; McClements, D.J. Improving the efficacy of essential oils as antimicrobials in foods: Mechanisms of action. *Annu. Rev. Food Sci. Technol.* **2019**, *10*, 365–387. [CrossRef]
20. Buendía-Moreno, L.; Ros-Chumillas, M.; Navarro-Segura, L.; Sánchez-Martínez, M.J.; Soto-Jover, S.; Antolinos, V.; Martínez-Hernández, G.B.; López-Gómez, A. Effects of an active cardboard box using encapsulated essential oils on the tomato shelf life. *Food Bioprocess Technol.* **2019**, *12*, 1548–1558. [CrossRef]
21. Kfoury, M.; Landy, D.; Fourmentin, S. Characterization of cyclodextrin/volatile inclusion complexes: A review. *Molecules* **2018**, *23*, 1204. [CrossRef]

22. Mortensen, A.; Aguilar, F.; Crebelli, R.; Di Domenico, A.; Dusemund, B.; Frutos, M.J.; Galtier, P.; Gott, D.; Gundert-Remy, U.; Leblanc, J.; et al. Re-evaluation of β-cyclodextrin (E 459) as a food additive. *EFSA J.* **2016**, *14*, e04628.

23. Khaneghah, A.M.; Hashemi, S.M.B.; Limbo, S. Antimicrobial agents and packaging systems in antimicrobial active food packaging: An overview of approaches and interactions. *Food Bioprod. Process.* **2018**, *111*, 1–19. [CrossRef]

24. Buendía-Moreno, L.; Soto-Jover, S.; Ros-Chumillas, M.; Antolinos, V.; Navarro-Segura, L.; Sánchez-Martínez, M.J.; Martínez-Hernández, G.B.; López-Gómez, A. Innovative cardboard active packaging with a coating including encapsulated essential oils to extend cherry tomato shelf life. *LWT* **2019**, *116*, 108584. [CrossRef]

25. Buendía-Moreno, L.; Sánchez-Martínez, M.J.; Antolinos, V.; Ros-Chumillas, M.; Navarro-Segura, L.; Soto-Jover, S.; Martínez-Hernández, G.B.; López-Gómez, A. Active cardboard box with a coating including essential oils entrapped within cyclodextrins and/or halloysite nanotubes. A case study for fresh tomato storage. *Food Control* **2020**, *107*, 106763. [CrossRef]

26. Kotronia, M.; Kavetsou, E.; Loupassaki, S.; Kikionis, S.; Vouyiouka, S.; Detsi, A. Encapsulation of oregano (*Origanum onites* L.) essential oil in β-Cyclodextrin (β-CD): Synthesis and characterization of the inclusion complexes. *Bioengineering* **2017**, *4*, 74. [CrossRef] [PubMed]

27. The European Commission. Commission Regulation Regulation (EU) No 1935/2004 of 14 January 2011 on materials and articles intended to come into contact with food and repealing Directives 80/590/EEC and 89/109/EEC. *Off. J. Eur. Union* **2004**, *338*, 4–17.

28. López-Gómez, A.; Ros-Chumillas, M.; Antolinos, V.; Buendía-Moreno, L.; Navarro-Segura, L.; Sánchez-Martínez, M.J.; Martínez-Hernández, G.B.; Soto-Jover, S. Fresh culinary herbs decontamination with essential oil vapours applied under vacuum conditions. *Postharvest Biol. Technol.* **2019**, *156*, 110942. [CrossRef]

29. Manolikar, M.; Sawant, M. Study of solubility of isoproturon by its complexation with β-cyclodextrin. *Chemosphere* **2003**, *51*, 811–816. [CrossRef]

30. Pathare, P.B.; Opara, U.L.; Al-Said, F.A.-J. Colour measurement and analysis in fresh and processed foods: A review. *Food Bioprocess Technol.* **2013**, *6*, 36–60. [CrossRef]

31. ASTM. *Physical Requirement Guidelines for Sensory Evaluation Laboratories*; Eggert, J., Zook, K., Eds.; ASTM International: Philadelphia, PA, USA, 1986; ISBN 0803109245.

32. ISO. *ISO 8589:2007: Sensory Analysis—General Guidance for the Design of Test Rooms*; ISO: Geneva, Switzerland, 2007.

33. D'hallewin, G.; Arras, G.; Castia, T.; Piga, A. Reducing decay of avana mandarin fruit by the use of uv, heat and thiabendazole treatments. *Acta Hortic.* **1994**, *368*, 387–394. [CrossRef]

34. Thompson, J.; Mitchell, F.G.; Kasmire, R.F. Cooling horticultural commodities. In *Postharvest Technology of Horticultural Crops*; University of California, Agriculture and Natural Resources: San Diego, CA, USA, 2002; pp. 111–130. ISBN 9781879906518.

35. Zhang, K.; Deng, Y.; Fu, H.; Weng, Q. Effects of Co-60 gamma-irradiation and refrigerated storage on the quality of Shatang mandarin. *Food Sci. Hum. Wellness* **2014**, *3*, 9–15. [CrossRef]

36. Pérez-Gago, M.B.; Rojas, C.; Del Río, M.A. Effect of hydroxypropyl methylcellulose-beeswax edible composite coatings on postharvest quality of "fortune" mandarins. In Proceedings of the Acta Horticulturae, International Society for Horticultural Science, Leuven, Belgium, 28 February 2003; Volume 599, pp. 583–587.

37. Ben-Yehoshua, S.; Burg, S.P.; Young, R. Resistance of citrus fruit to mass transport of water vapor and other gases. *Plant Physiol.* **1985**, *79*, 1048–1053. [CrossRef]

38. Leide, J.; Hildebrandt, U.; Reussing, K.; Riederer, M.; Vogg, G. The developmental pattern of tomato fruit wax accumulation and its impact on cuticular transpiration barrier properties: Effects of a deficiency in a β-ketoacyl-coenzyme a synthase (LeCER6). *Plant Physiol.* **2007**, *144*, 1667. [CrossRef]

39. Choi, W.S.; Singh, S.; Lee, Y.S. Characterization of edible film containing essential oils in hydroxypropyl methylcellulose and its effect on quality attributes of "Formosa" plum (*Prunus salicina* L.). *LWT Food Sci. Technol.* **2016**, *70*, 213–222. [CrossRef]

40. Rapisarda, P.; Bianco, M.L.; Pannuzzo, P.; Timpanaro, N. Effect of cold storage on vitamin C, phenolics and antioxidant activity of five orange genotypes [*Citrus sinensis* (L.) Osbeck]. *Postharvest Biol. Technol.* **2008**, *49*, 348–354. [CrossRef]

41. Katz, E.; Lagunes, P.; Riov, J.; Weiss, D.; Goldschmidt, E.E. Molecular and physiological evidence suggests the existence of a system II-like pathway of ethylene production in non-climacteric Citrus fruit. *Planta* **2004**, *219*, 243–252. [CrossRef] [PubMed]

42. Kusunose, H.; Sawamura, M. Ethylene production and respiration of postharvest acid citrus fruits and Wase satsuma mandarin fruit. *Agric. Biol. Chem.* **1980**, *44*, 1917–1922. [CrossRef]

43. Rabbany, A.B.M.G.; Mizutani, F. Effect of essential oils on ethylene production and ACC content in apple fruit and peach seed tissues. *Engei Gakkai Zasshi* **1996**, *65*, 7–13. [CrossRef]

44. Zapata, P.J.; Castillo, S.; Valero, D.; Guillén, F.; Serrano, M.; Díaz-Mula, H.M. The use of alginate as edible coating alone or in combination with essential oils maintained postharvest quality of tomato. *Acta Hortic.* **2010**, *877*, 1529–1534. [CrossRef]

45. Pérez-Alfonso, C.O.; Martínez-Romero, D.; Zapata, P.J.; Serrano, M.; Valero, D.; Castillo, S. The effects of essential oils carvacrol and thymol on growth of *Penicillium digitatum* and *P. italicum* involved in lemon decay. *Int. J. Food Microbiol.* **2012**, *158*, 101–106. [CrossRef] [PubMed]

46. Alhassan, N.; Golding, J.B.; Wills, R.B.H.; Bowyer, M.C.; Pristijono, P. Long term exposure to low ethylene and storage temperatures delays calyx senescence and maintains 'Afourer' mandarins and navel oranges quality. *Foods* **2019**, *8*, 19. [CrossRef]

47. Valverde, J.M.; Guillén, F.; Martínez-Romero, D.; Castillo, S.; Serrano, M.; Valero, D. Improvement of table grapes quality and safety by the combination of modified atmosphere packaging (MAP) and eugenol, menthol, or thymol. *J. Agric. Food Chem.* **2005**, *53*, 7458–7464. [CrossRef]

48. Martínez-Romero, D.; Guillén, F.; Valverde, J.M.; Bailén, G.; Zapata, P.; Serrano, M.; Castillo, S.; Valero, D. Influence of carvacrol on survival of Botrytis cinerea inoculated in table grapes. *Int. J. Food Microbiol.* **2007**, *115*, 144–148. [CrossRef]

49. Chen, X.; Ren, L.; Li, M.; Qian, J.; Fan, J.; Du, B. Effects of clove essential oil and eugenol on quality and browning control of fresh-cut lettuce. *Food Chem.* **2017**, *214*, 432–439. [CrossRef]

50. Echeverria, E.; Ismail, M. Changes in sugar and acids of citrus fruits during storage. *Proc. Fla. State Hortic. Soc.* **1987**, *100*, 50–52.

51. Obenland, D.; Collin, S.; Mackey, B.; Sievert, J.; Arpaia, M.L. Storage temperature and time influences sensory quality of mandarins by altering soluble solids, acidity and aroma volatile composition. *Postharvest Biol. Technol.* **2011**, *59*, 187–193. [CrossRef]

52. Campbell, B.L.; Nelson, R.G.; Ebel, R.C.; Dozier, W.A.; Campbell, J.H.; Woods, F.M. Mandarin market segments based on consumer sensory evaluations. *J. Food Distrib. Res.* **2008**, *39*, 43–55.

53. Jaya, S.; Das, H. Sensory evaluation of mango drinks using fuzzy logic. *J. Sens. Stud.* **2003**, *18*, 163–176. [CrossRef]

54. Alquézar, B.; Rodrigo, M.J.; Zacarías, L. Carotenoid biosynthesis and its regulation in citrus fruits. *Tree Sci. Biotechnol.* **2008**, *2*, 23–35.

55. Reid, M.S. Ethylene in postharvest technology. In *Postharvest Technology of Horticultural Crops*; Kader, A.A., Ed.; University of California, Agriculture and Natural Resources: San Diego, CA, USA, 2002; p. 149. ISBN 9781879906518.

56. Tournas, V.H.; Katsoudas, E. Mould and yeast flora in fresh berries, grapes and citrus fruits. *Int. J. Food Microbiol.* **2005**, *105*, 11–17. [CrossRef]

57. Ren, X.; Yue, S.; Xiang, H.; Xie, M. Inclusion complexes of eucalyptus essential oil with β-cyclodextrin: Preparation, characterization and controlled release. *J. Porous Mater.* **2018**, *25*, 1577–1586. [CrossRef]

58. Boskovic, M.; Glisic, M.; Djordjevic, J.; Starcevic, M.; Glamoclija, N.; Djordjevic, V.; Baltic, M.Z. Antioxidative activity of thyme (*Thymus vulgaris*) and oregano (*Origanum vulgare*) essential oils and their effect on oxidative stability of minced pork packaged under vacuum and modified atmosphere. *J. Food Sci.* **2019**, *84*, 2467–2474. [CrossRef]

59. Zhu, P.; Xu, L.; Zhang, C.; Toyoda, H.; Gan, S.-S. Ethylene produced by *Botrytis cinerea* can affect early fungal development and can be used as a marker for infection during storage of grapes. *Postharvest Biol. Technol.* **2012**, *66*, 23–29. [CrossRef]

© 2020 by the authors. Licensee MDPI, Basel, Switzerland. This article is an open access article distributed under the terms and conditions of the Creative Commons Attribution (CC BY) license (http://creativecommons.org/licenses/by/4.0/).

 *foods*

Article

Combined Effect of Dipping in Oxalic or in Citric Acid and Low O$_2$ Modified Atmosphere, to Preserve the Quality of Fresh-Cut Lettuce during Storage

Bernardo Pace [1,*], Imperatrice Capotorto [1], Michela Palumbo [1,2], Sergio Pelosi [1] and Maria Cefola [1,*] 

[1]   Institute of Sciences of Food Production, National Research Council of Italy (CNR), c/o CS-DAT, Via Michele Protano, 71121 Foggia, Italy; imperatrice.capotorto@ispa.cnr.it (I.C.); michela.palumbo@ispa.cnr.it (M.P.); sergio.pelosi@ispa.cnr.it (S.P.)

[2]   Department of Science of Agriculture, Food and Environment, University of Foggia, Via Napoli 25, 71122 Foggia, Italy

*   Correspondence: bernardo.pace@ispa.cnr.it (B.P.); maria.cefola@ispa.cnr.it (M.C.); Tel.: +39-0881-630201 (B.P.)

Received: 30 June 2020; Accepted: 21 July 2020; Published: 24 July 2020

Abstract: Leaf edge browning is the main factor affecting fresh-cut lettuce marketability. Dipping in organic acids as well as the low O$_2$ modified atmosphere packaging (MAP), can be used as anti-browning technologies. In the present research paper, the proper oxalic acid (OA) concentration, able to reduce respiration rate of fresh-cut iceberg lettuce, and the suitable packaging materials aimed to maintaining a low O$_2$ during storage, were selected. Moreover, the combined effect of dipping (in OA or in citric acid) and packaging in low O$_2$ was investigated during the storage of fresh-cut iceberg lettuce for 14 days. Results showed a significant effect of 5 mM OA on respiration rate delay. In addition, polypropylene/polyamide (PP/PA) was select as the most suitable packaging material to be used in low O$_2$ MAP. Combining OA dipping with low O$_2$ MAP using PP/PA as material, resulted able to reduce leaf edge browning, respiration rate, weight loss and electrolyte leakage, preserving the visual quality of fresh-cut lettuce until 8 days at 8 °C.

Keywords: *Lactuca sativa* L.; minimally processed lettuce; modified atmosphere packaging; oxalic acid; shelf life

---

1. Introduction

Iceberg lettuce (*Lactuca sativa* L.) is considered one of the most popular vegetable and represent the primary fresh-cut product processed in Italy as well as in many other European and North America countries. Iceberg lettuce is highly desired by worldwide consumers and processors for its sensory and technological properties [1]. When lettuce is processed as fresh-cut product, an accelerated metabolism, due to the processing operations, causes, among the other alterations, the browning of cut surface [1–4]. Peroxidase (POD; EC 1.11.1.7) and polyphenol oxidase (PPO; EC 1.14.18.1) are enzymes that catalyze the conversion of polyphenols into quinones. In particular PPO catalyzes both hydroxylation of monophenols to o-diphenols and oxidation of colorless o-diphenols to o-quinones give brown pigments to cut tissues. While POD promote the oxidation of phenols to quinones in the presence of hydrogen peroxide [1]. To limit these changes that affect quality, supplementary techniques have been used in addition to the low temperature, such as modified atmosphere, chemical dipping based on different organic acids, edible coatings, heat treatments [1,5–7]. Saltveit [8] reports that modified atmosphere packaging (MAP) with low O$_2$ and high CO$_2$ was successfully applied to control browning specifically for the iceberg lettuce, whereas Gorny [9] shows that atmospheres with O$_2$ values between 0.5 kPa and 3 kPa are recommended for iceberg lettuce. In addition, natural dipping in solutions based on ascorbic, citric, salicylic or oxalic acid, to prevent leaf edge browning can be used. These substances are recognized as safe for food applications and consumption, and are widely used for fresh-cut products [7,10,11]. Citric acid has been extensively used for its anti-browning activity in minimally processed fruits and vegetables [12] and specifically on fresh lettuce [5]. Moreover, it was also reported the efficiency of oxalic acid in preventing lettuce browning and extending the shelf life of fresh-cut fruits and vegetables [5,10,11,13–16].

Starting from these considerations, the goal of the present work is to study the combined effect of dipping in oxalic or citric acid and low O_2 MAP to control the leaf edge browning of fresh-cut lettuce. To this aim, the proper oxalic acid concentration, and the suitable packaging material able to maintain an adequate low O_2 concentration inside MAP bags were selected in specifically preliminary trials. Changes in sensory characteristics, respiration rate and chemical parameters (such as ammonium content, electrolyte leakage, phenolic content and *o*-quinones) during cold storage were investigated. To our knowledge, this is the first manuscript describing the effect of oxalic acid on the quality of fresh-cut iceberg lettuce.

2. Materials and Methods

2.1. Plant Material and Sample Preparation

Iceberg lettuce heads (*Lactuca sativa* L.) were obtained from a local farm located in Foggia, at the same growing stage and were transported to the laboratory under refrigerated conditions in polystyrene boxes. Samples were kept in darkness for 24 h at 4 °C and at 85% relative humidity before being processed.

Lettuce heads were processed by removing outer leaves and the stem with stainless steel knives, then they were cut using a vegetable cutter (CL52 Robot Coupe, Vincennes-Cedex, France). Thus, the fresh-cut iceberg lettuce pieces obtained (3 × 4 cm) were pooled and blended, to minimize product heterogeneity, and were used for the experiments described below.

Two preliminary trials, the first aimed to select the proper oxalic acid (OA) concentration and the other finalized to the choice of the suitable packaging material to be used in the MAP, were carried out.

Then, based on results of preliminary trials, an experiment was conducted in order to show the combined effect of dipping in OA or citric acid and low O_2 MAP on fresh-cut iceberg lettuce.

2.1.1. Preliminarily Trials

Selection of the Proper Oxalic Acid Concentration

Four lots of fresh-cut iceberg lettuce, obtained as reported above, were used. Each lot was washed, firstly, in tap water for 5 min at 5 °C, and subsequently dipped for 1 min at 5 °C in one of the three different OA concentrations: 1, 3, or 5 mM OA. Samples dipped in tap water were used as control (CTRL). After dipping, lettuce pieces belonging to each dipping treatment were immediately placed in a manual centrifuge and dried for about 1 min to remove the water excess. Then, lettuce pieces were placed in open polypropylene (PP) bags. A total of 36 bags (3 replicate × 4 treatments, 1 mM OA, 3 mM OA, 5 mM OA, or CTRL, × 3 storage time, 2, 4, and 7 days) were prepared. All bags, with about 150 g of fresh-cut lettuce each, were stored at 8 (±1) °C. At 0 days and at each storage time, all samples were analyzed for the respiration rate.

Selection of the Proper Packaging Material

In the second preliminary trial, three packaging materials were compared: polypropylene (PP, 30 μm thickness, OTR 1100 $cm^3 \cdot m^{-2}$ 24 $h^{-1} \cdot bar^{-1}$), polypropylene/polyamide (PP/PA, 67 μm thickness, OTR 100 $cm^3 \cdot m^{-2}$ 24 $h^{-1} \cdot bar^{-1}$), and PP/PA micro-perforated (PP/PA MP, 67 μm thickness) (Carton Pack Srl, Rutigliano, Italy).

Fresh-cut lettuce was dipped for 1 min at 5 °C in tap water and then centrifuged for about 1 min to remove water excess. In each bag (21 × 18 cm), 150 g of fresh-cut iceberg lettuce pieces were placed. For each packaging material (PP, PP/PA or PP/PA MP) 12 bags were prepared and sealed using a packaging machine (Boxer 50 GAS-Lavezzini, Fiorenzuola d'Arda, Italy) with an initial atmosphere of 3 kPa O_2 and 97 kPa N_2.

In addition, 12 open PP bags with 150 g of fresh-cut iceberg lettuce each were used as control (CTRL). All bags were stored at 8 °C for 6 days. Initially, and at each storage time (1, 2, 3 and 6 days), the visual quality of lettuce pieces was evaluated and the atmosphere composition inside packages was analysed.

2.1.2. Combined Effect of Dipping in Oxalic or in Citric Acid and Low O_2 MAP

The results of the preliminary trials were applied in the present experiment. Fresh-cut lettuce was washed in tap water for 5 min at 5 °C, and then dipped for 1 min at 5 °C in one of the following different solutions: 5 mM OA (the best concentration resulted from preliminary trial), or 1% citric acid (w:v) (CIT). Samples dipped in tap water in the same conditions, were used as control (CTRL). After dipping, lettuce pieces, belonging to each treatment (OA, CIT or CTRL) were immediately placed in a manual centrifuge for about 1 min to remove water excess. Dipped samples (OA, CIT or CTRL) were subsequently placed in PP/PA bags (21 × 18 cm) (the proper material resulted from preliminary trial) and sealed, with an initial modified atmosphere (MAP) of 3 kPa O_2 and 97 kPa N_2, using the packaging machine.

A total of 27 bags (3 replicates × 3 dipping treatments, OA, CIT, CTRL x 3 storage time, after 3, 6 and 8 days) were prepared. All bags, with about 150 g fresh-cut iceberg lettuce each, were stored at 8 (±1) °C. The quality parameters measured, initially, and at each storage time, are described below.

2.2. Respiration Rate and Headspace Analysis

The respiration rate of fresh-cut iceberg lettuce was measured at 8 °C using a closed system as previously described by Kader [17] initially and during storage (just after the opening of the bags for MAP samples). About 100 g of product, for each treatment and replicate ($n = 3$), was placed into 6 L sealed plastic jars (one jar per replicate) where CO_2 was allowed to accumulate up to 0.1 kPa (concentration of the CO_2 standard). The CO_2 analysis was conducted taking 1 mL of gas sample from the headspace of the plastic jars and injecting into a gas chromatograph (p200 micro GC-Agilent, Santa Clara, CA, USA) equipped with dual columns and a thermal conductivity detector. Carbon dioxide was analyzed with a retention time of 16 s and a total run time of 120 s on a 10 m porous polymer (PPU) column (Agilent, Santa Clara, CA, USA) at a constant temperature of 70 °C. Respiration rate was expressed as mL CO_2 $kg^{-1} \cdot h^{-1}$.

Gas composition (O_2 and CO_2 kPa) of packages in MAP was measured during storage, using a gas analyser (CheckPoint O_2/CO_2 Dansensor® Mocon, Ringsted, Denmark).

2.3. Sensory Evaluation, Colour, Texture and Weight Loss

Fresh-cut iceberg lettuce belonging to each treatment was examined by a group of eight trained researchers at the beginning of the experiment (on the fresh lettuce) and at each storage time. Coded (3 digits) samples were presented to the trained researchers (judges) individually, to enable them to make independent evaluations.

Visual quality was evaluated on a 5-point rating scale according to Cefola et al. [18] where 5: excellent (fresh appearance, full sensory acceptability); 4: good (product acceptable from a sensory point of view); 3: limit of sensory acceptability; 2: product has notable visual defects; 1: severe visual defects. Samples rated below 3 were considered as unmarketable for the loss of the sensory visual quality (loss of turgor and brightness accompanied by softening and browning of leaf tissues). A color photographic scale, accompanying with a brief description of freshness, color uniformity, and brightness, was applied as reference.

The color of lettuce pieces was measured using a colorimeter (CR-400-Konica Minolta, Osaka, Japan) equipped with a D65 illuminant in the reflectance mode and in the CIE L^* a^* b^* color scale. In detail, L^* indicates the lightness from black (0 value) to white (100 value), a^* the redness (+) or greenness (-), and b^* the yellowness (+) or blueness (-).

The color was measured on three random points on the midribs surface of 5 pieces of lettuce for each replicate. The instrument was calibrated with a white plate as standard reference ($L^* = 97.55$, $a^* = 1.32$, $b^* = 1.41$). The a^* and b^* color parameters recorded, were used for the calculation of the hue angle (h°) using the following formula [19].

$$h^\circ = arctg\frac{b*}{a*},\tag{1}$$

where, a^* and b^* are the color parameters acquired by colorimeter.

The texture of the fresh-cut lettuce was assessed using a texture analyzer (ZwickLine Z0.5-Zwick/Roell, Ulm, Germany) equipped with a Kramer shear cell with 10 blades. For each replicate 5 samples of about 3 g each one were used. Texture was expressed as the maximum peak force (N) per gram of lettuce (N g^{-1}).

The weight loss of each replicate was calculated as percentage compared with the initial weight.

2.4. Electrolyte Leakage and Ammonium Content

Electrolyte leakage was measured applying the procedure described by Kim et al. [20] with slight modifications. For each replicate, about 2.5 g of lettuce disks of 8 mm of diameter, obtained using a cork borer, were immersed in tubes containing 25 mL of distilled water. After 30 min of storage at 8 °C, the conductivity of the solution was measured by a conductivity meter (Cond. 51+, XS Instruments, Carpi, Italy). The tubes, with the vegetable portion, were then frozen and after 24 h, samples were thawed, and the total conductivity was measured. Electrolyte leakage was calculated as the percentage ratio of initial over total conductivity.

Ammonium content was evaluated according to Fadda et al. [21]. Lettuce pieces (5 g) were chopped and homogenized (T 25 digital Ultra-Turrax - IKA, Staufen, Germany) for 2 min in 20 mL of distilled water on an ice bath, and then centrifuged (Prism R C2500-R - Labnet, Edison, NJ, USA) for 5 min at 6440× g at 4 °C. Then, the supernatant extract (0.5 mL), was mixed with 5 mL of nitroprusside reagent (phenol and hypochlorite in alkali reaction mixture) and heated at 37 °C for 20 min. The color development after incubation, was determined with a spectrophotometer (UV-1800 - Shimadzu, Kyoto, Japan) reading the absorbance at 635 nm. The content of NH_4^+ was expressed as mmol NH_4^+ kg^{-1} of fresh weight, using ammonium sulfate as standard (0–10 µg mL^{-1}, $R^2 = 0.99$).

2.5. Total Phenol Content and O-Quinones Determination

The total phenol content was determined according to the method of Fadda et al. [21]. Five grams of chopped lettuce for each replication was homogenized in a methanol:water solution (80:20) for 1 min and then centrifuged at 4 °C at 6440× g for 5 min. The absorbance was read after 2 h at 765 nm. The total phenol content was calculated based on the calibration curve of gallic acid and was expressed as milligrams of gallic acid per kg of fresh weight.

Soluble *o*-quinones of lettuce tissues, were extracted as described by Degl'Innocenti et al. [22] with low modifications. Five grams of tissues were homogenized with 20 mL methanol:water solution (80:20) for 1 min, filtered and centrifuged at 4 °C at 6440× g for 5 min. The supernatant was used directly to measure the soluble *o*-quinones at a wavelength of 437 nm. The result was expressed as the absorbance for 5 g of fresh weight.

2.6. Statistical Analysis

For each preliminary trial, a two-way ANOVA for $p \leq 0.05$ was performed to evaluate the effects of treatments, storage time and their interaction on quality parameters measured. As for the main experiment, a two-way ANOVA for $p \leq 0.05$ was performed to evaluate the effects of combined treatments (dipping and packaging), storage time and their interaction on quality parameters. When the interaction between factors was significant, data were shown as graphs with mean values ± standard deviation. The statistical analysis was performed using the software STATISTICA 6.0 (StatSoft, Hamburg, Germany).

3. Results

3.1. Oxalic Acid Concentration and Packaging Selection

In the first preliminary trial, three different OA concentrations were compared (1 mM, 3 mM or 5 mM) as fresh-cut lettuce dipping treatments for 1 min at 5 °C and the respiration rate was monitored during storage in air at 8 °C. Respiration rate was significantly affected by interaction between dipping treatments and storage time ($p \leq 0.001$). Fresh-cut lettuces dipped in OA showed the lowest respiration rate during the entire storage (Figure 1). In particular, after 2 days of storage in air at 8 °C, fresh-cut lettuce dipped in 5 mM OA showed a respiration rate about 50% lower than CTRL. Then, after 4 and 7 days, no significant differences compared to CTRL were measured. A similar trend was also detected for the other two dipping solutions (1 or 3 mM), but in these cases the reduction in respiration rate respect to CTRL after 2 days of storage was about 22% and 38% for 1 and 3 mM OA, respectively.

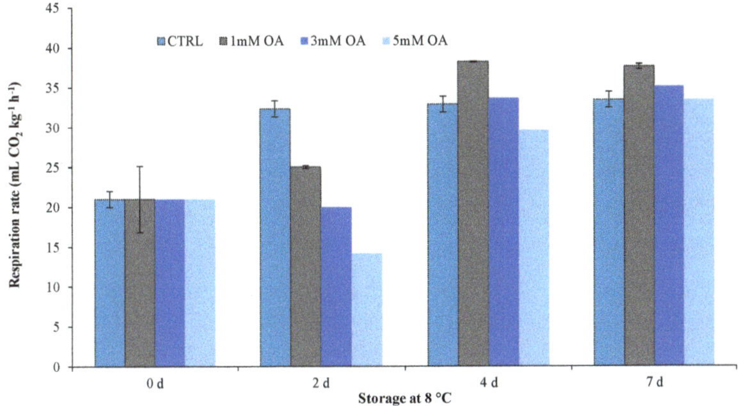

Figure 1. Respiration rate measured during the storage at 8 °C in fresh-cut iceberg lettuce dipped with different oxalic acid (OA) concentrations respect to control samples (CTRL).

Concerning the second preliminary trial, Figure 2 shows the changes in O_2 and CO_2 inside the bags closed in MAP with 3 kPa O_2 and 97 kPa N_2. In PP and PP/PA MP bags the O_2 and CO_2 increased during storage until to values between 5 and 10 kPa for both gases. On the other hand, in PP/PA bags the O_2 concentration decreased until 1 kPa after 1 day of storage, then this value remained constant until the end of the trial, while the CO_2 concentration increased reaching, at the end of the storage, a value of over 15 kPa. The gas composition measured inside PP/PA bags, positively affected the fresh-cut lettuce visual quality (Figure 3). Indeed, fresh-cut lettuce stored in PP/PA showed the highest visual quality score during the entire storage, followed by samples stored in PP, PP/PA MP and CTRL. Moreover, lettuce stored in air lost marketability after only 2 days of storage, mainly for the severe browning of the cut-surface. On the other hand, samples stored in MAP resulted marketable for 4–6 days in PP or PP/PA MP bags, while a long marketability (more than 6 days) might be reached when fresh-cut lettuce was packed in PP/PA. To sum up, the results of these two preliminary trials were used in the main experiment aimed to select the proper combination of dipping and MAP to improve fresh-cut lettuce marketability.

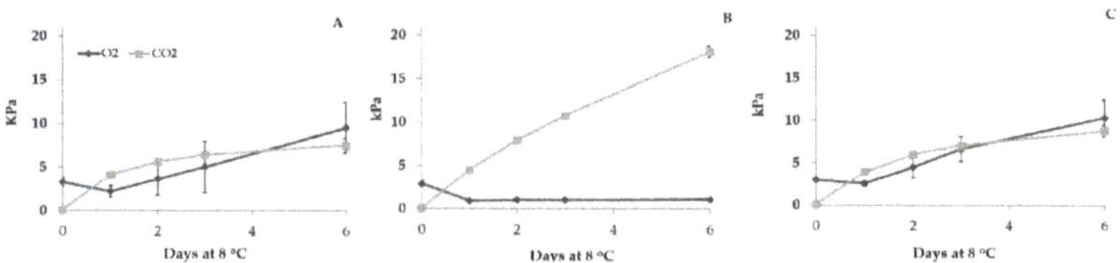

Figure 2. Changes in O_2 and CO_2 inside PP (**A**), PP/PA (**B**), PP/PA MP (**C**) fresh-cut iceberg lettuce bags. Data are mean ($n = 3$) ± standard deviation.

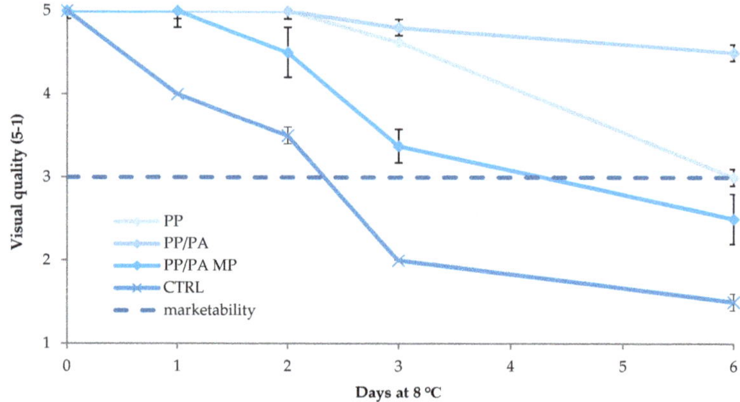

Figure 3. Changes in visual quality of fresh-cut iceberg lettuce stored in MAP using PP, PP/PA or PP/PA MP bags or in air (CTRL). Data are mean ($n = 3$) ± standard deviation.

3.2. Combined Effect of Dipping and MAP to Extend the Marketability of Fresh-Cut Lettuce

Results of the two-way ANOVA showed a significant effect of interaction (combined treatment × storage) on visual quality, hue angle and respiration rate (Table 1). Total phenol and *o*-quinones were affected by storage and combined treatment, respectively, while weight loss, electrolyte leakage and ammonium were affected by both factors (Table 1). The main effect of each factor on these parameters (weight loss, electrolyte leakage, ammonium content, total phenols and *o*-quinones) are reported in Table 2. The product treated with OA-MAP showed, a significant lower mean value for weight loss, electrolyte leakage and o-quinones than the other combined treatments (Table 2).

The highest value of ammonium content was measured for CIT-MAP, whereas no differences on total phenol content was reported among treatments (Table 2).

Nonetheless, during storage, a significant increase in weight loss, ammonium content, and total phenols was measured in all samples, while for *o*-quinones no significant difference was observed (Table 2).

In Figure 4, changes in VQ, hue angle and respiration rate during storage are reported. In particular, the VQ value, remained almost constant in sample OA-MAP, while a slight reduction was observed for CIT-MAP and CTRL-MAP after 3 and 6 days, respectively, although the mean values remained above the marketable limit (VQ = 3) for the entire trial in all samples (Figure 4A). As for h°, the initial value remained almost constant during storage for all samples and only in CTRL-MAP sample a slight reduction after 3 days of storage was measured (Figure 4B).

Starting from an initial value of about 15 mL CO_2 kg^{-1} h^{-1}, respiration rate increased in all samples over time: in details, OA-MAP and CIT-MAP showed a similar trend, with an increase at the end of the trial of about 30% compared to fresh samples, while the rise in CTRL-MAP was of roughly 45% (Figure 4C).

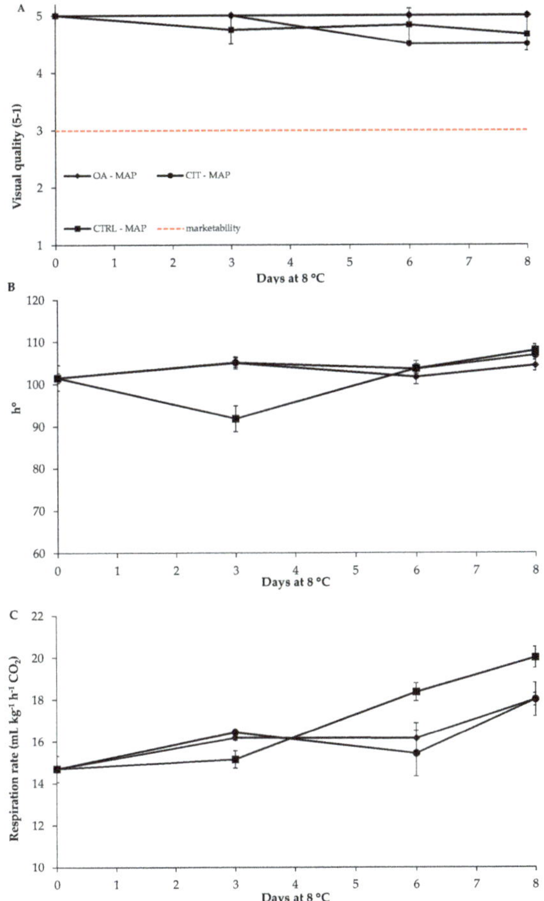

Figure 4. Changes in visual quality (**A**), hue angle (**B**), and respiration rate (**C**) in fresh-cut iceberg lettuce dipped in oxalic acid (OA), citric acid (CIT) or control (CTRL) and stored in modified atmosphere packaging (MAP). Data are mean value ($n = 3$) $\pm$ standard deviation.

Table 1. Effect of combined treatments (dipping in oxalic acid, citric acid or water and packaging in MAP) (A), storage (3, 6, 8 days at 8 °C) (B), and their interaction (A × B) on quality parameters of fresh-cut iceberg lettuce.

Parameter	Combined Treatment (A)	Storage (B)	Combined Treatment × Storage (A × B)
Respiration rate (mL CO_2 kg^{-1}·h^{-1})	ns	*	***
Hue angle (h°)	***	***	***
Texture (N·g^{-1})	ns	ns	ns
Weight loss %	*	***	ns
Visual Quality (VQ)	***	ns	*
Off Odour	ns	ns	ns
Electrolyte leakage %	*	*	ns
Ammonium content (mmol·kg^{-1} NH$_4^+$)	**	***	ns
Total phenols (mg·kg^{-1} gallic acid)	ns	*	ns
o-quinones (OD 427 nm)	*	ns	ns

Asterisks indicate the significance level for each factor of the ANOVA test (ns, not significant; * $p \leq 0.05$; ** $p \leq 0.01$; *** $p \leq 0.0001$).

Table 2. Main effects of main factors (combined treatments or storage) on the postharvest quality of fresh-cut iceberg lettuce.

Main Factors	Weight Loss %		Electrolyte Leakage %		Ammonium Content (mmol·kg^{-1} NH$_4{}^+$)		Total Phenols (mg·kg^{-1} gallic acid)		O-Quinones (OD 427 nm)	
					Combined Treatments					
OA-MAP	0.361	b	41.4	b	1.99	b	83.1	ns	0.10	b
CIT-MAP	0.445	ab	44.3	a	2.17	a	89.6	ns	0.12	ab
CTRL-MAP	0.512	a	44.1	a	1.96	b	83.9	ns	0.14	a
Storage										
3	0.295	b	42.5	b	1.72	b	76.4	b	0.13	ns
6	0.484	a	45.2	a	2.15	a	88.6	a	0.11	ns
8	0.539	a	42.1	b	2.27	a	91.7	a	0.13	ns

For each factor (combined treatments or storage) different letters (a,b) indicate significant differences ($p < 0.05$) according to Dunkan's test. ns, not significant.

4. Discussion

Fresh-cut lettuce marketability is mainly affected by the leaf edge browning, due to the action of polyphenol oxidase (PPO) on phenols [23,24]. As a consequence, the shelf life of fresh-cut lettuce is limited to 2–3 days at 8 °C in air. In this research paper, the combined effect of dipping in oxalic acid or citric acid and MAP (3 kPa O_2 in 97 kPa N_2) was studied with the goal to extend the fresh-cut lettuce marketability.

The low PP/PA permeability to O_2 and CO_2 at 8 °C, promoted inside packages a flow of O_2 adequate for product respiration. Thus, after a slight reduction from the initial value, the O_2 concentration remained stable at 1 kPa, while CO_2 increased to about 15 kPa. It was reported that O_2 concentrations below 3 kPa, but not less than 0.5–1 kPa, avoid the fermentation and browning of the cut surface in fresh-cut lettuce [25]. These atmosphere conditions positively affected the storability of fresh-cut lettuce as previously observed [26]. Moreover, these concentrations are suitable to delay browning in fresh-cut lettuce due to the reduction in the phenolic metabolism [27–29].

The combined treatment based on dipping in OA and low O_2 MAP allowed controlling respiration rate and weight loss, limiting the browning development and electrolyte leakage, finally preserving the visual quality. OA is generally recognized as safe (GRAS) compound, useful in pre and postharvest treatments on fruits and vegetables [15]. The positive effect of OA concentration applied on respiration rate and weight loss, might be due to a delay of metabolic activity, as reported for artichoke [30], asparagus [13], and rocket and baby spinach leaves [11]. In addition, Cefola et al. [11] reported that OA dipping might act as antioxidant controlling vegetable tissue browning as well as preserving the leaf membrane integrity, limiting the increase in electrolytic leakage during storage [16]. It was reported that OA might directly inhibit the activities of the enzymes involved in browning reactions (PPO, POD), reducing the pH of the treated produce. On the other hand, OA could react with soluble quinones, reducing them to into uncoloured catechol or creating colourless adducts [6,31].

5. Conclusions

Oxalic acid 5 mM was selected as proper concentration able to delay the respiration rate of fresh-cut iceberg lettuce. Moreover, suitable packaging was obtained using polypropylene/polyamide bags closed in modified atmosphere packaging with 3 kPa O_2 in 97 kPa N_2. Combining dipping in oxalic acid or citric acid with low O_2 modified atmosphere packaging, the best results were obtained using oxalic acid as treatment. In particular, the combined effect of oxalic acid dipping and low O_2 modified atmosphere packaging allowed us to reduce the respiration rate, weight loss, and electrolyte leakage during storage, preserving the visual quality of fresh-cut iceberg lettuce stored at 8 °C until 8 days.

Author Contributions: Conceptualization, B.P. and M.C.; methodology, B.P. and M.C.; validation, B.P. and M.C.; formal analysis, I.C., M.P. and S.P.; investigation, B.P., M.C., and M.P.; data curation, B.P., M.C. and M.P.; writing—original draft preparation, B.P., M.C., S.P. and I.C.; writing—review and editing, B.P. and M.C.; visualization, B.P., M.C. and I.C.; supervision, B.P. and M.C.; All authors have read and agreed to the published version of the manuscript.

Funding: This research received no external funding.

Acknowledgments: The authors thank Massimo Franchi and Salvatore Burbaci for providing technical assistance during the experiment.

Conflicts of Interest: The authors declare no conflict of interest.

References

1. Pace, B.; Capotorto, I.; Ventura, M.; Cefola, M. Evaluation of L-cysteine as anti-browning agent in fresh-cut lettuce processing. *J. Food Process. Preserv.* **2015**, *39*, 985–993. [CrossRef]

2. Heimdal, H.; Kühn, B.F.; Poll, L.; Larsen, L.M. Biochemical changes and sensory quality of shredded and MA-packaged iceberg lettuce. *J. Food Sci.* **1995**, *60*, 1265–1268. [CrossRef]

3. Amodio, M.L.; Cefola, M.; Pace, B.; Colelli, G. Fresh-Cut Fruits and Vegetables. In *Postharvest Physiological Disorders in Fruits and Vegetables*, 1st ed.; Tonetto de Freitas, S., Pareek, S., Eds.; CRC Press: Boca Raton, FL, USA, 2019; pp. 761–784. ISBN 13:978-1138035508.

4. Tudela, J.A.; Gil, M.I. Leafy vegetables: Fresh-cut lettuce. In *Controlled and Modified Atmospheres for Fresh and Fresh-Cut Produce*, 1st ed.; Gil, M.I., Beaudry, R., Eds.; Academic Press: Cambridge, MA, USA, 2020; pp. 545–550. [CrossRef]

5. Altunkaya, A.; Gökmen, V. Effect of various inhibitors on enzymatic browning, antioxidant activity and total phenol content of fresh lettuce (*Lactuca sativa*). *Food Chem.* **2008**, *107*, 1173–1179. [CrossRef]

6. Altunkaya, A.; Gökmen, V. Effect of various anti-browning agents on phenolic compounds profile of fresh lettuce (*L. sativa*). *Food Chem.* **2009**, *117*, 122–126. [CrossRef]

7. Soares, C.D.F.; Martin, J.G.P.; Berno, N.D.; Kluge, R.A. Antioxidant chemical treatment affects physiology and quality of minimally-processed escarole. *Horticulturae* **2019**, *5*, 75. [CrossRef]

8. Saltveit, M.E. Physical and physiological changes in minimally processed fruits and vegetables. In *Phytochemistry of Fruit and Vegetables*; Tomás-Barberán, F.A., Ed.; Oxford University Press: New York, NY, USA, 1997; pp. 205–220.

9. Gorny, J.R. A summary of CA and MA requirements and recommendations for fresh-cut (minimally processed) fruits and vegetables. *Acta Hort.* **2003**, *600*, 609–614. [CrossRef]

10. Suttirak, W.; Manurakchinakorn, S. Potential application of ascorbic acid, citric acid and oxalic acid for browning inhibition in fresh-cut fruits and vegetables. *Walailak J. Sci. Tech.* **2010**, *7*, 5–14.

11. Cefola, M.; Pace, B. Application of oxalic acid to preserve the overall quality of rocket and baby spinach leaves during storage. *J. Food Process. Preserv.* **2015**, *39*, 2523–2532. [CrossRef]

12. Ahvenainen, R. New approaches in improving the shelf life of minimally processed fruit and vegetables. *Trends Food Sci. Technol.* **1996**, *7*, 179–187. [CrossRef]

13. Barberis, A.; Cefola, M.; Pace, B.; Azara, E.; Spissu, Y.; Serra, P.R.; Logrieco, A.F.; D'hallewin, G.; Fadda, A. Postharvest application of oxalic acid to preserve overall appearance and nutritional quality of fresh-cut green and purple asparagus during cold storage: A combined electrochemical and mass-spectrometry analysis approach. *Postharvest Biol. Technol.* **2019**, *148*, 158–167. [CrossRef]

14. Zheng, J.; Li, S.; Xu, Y.; Zheng, X. Effect of oxalic acid on edible quality of bamboo shoots (Phyllostachys prominens) without sheaths during cold storage. *LWT* **2019**, *109*, 194–200. [CrossRef]

15. Ali, S.; Khan, A.S.; Anjum, M.A.; Nawaz, A.; Naz, S.; Ejaz, S.; Hussain, S. Effect of postharvest oxalic acid application on enzymatic browning and quality of lotus (Nelumbo nucifera Gaertn.) root slices. *Food Chem.* **2020**, *312*, 126051. [CrossRef] [PubMed]

16. Zheng, X.L.; Brecht, J.K. Oxalic acid treatments. In *Novel Postharvest Treatments of Fresh Produce*, 1st ed.; Pareek, S., Ed.; CRC Press: Boca Raton, FL, USA, 2018; pp. 35–49. [CrossRef]

17. Kader, A.A. Methods of gas mixing, sampling and analysis. In *Postharvest Technology of Horticultural Crops*, 3rd ed.; Kader, A.A., Ed.; ANR Publication: Oakland, CA, USA, 2020; pp. 145–148.

18. Cefola, M.; Carbone, V.; Minasi, P.; Pace, B. Phenolic profiles and postharvest quality changes of fresh-cut radicchio (*Cichorium intybus* L.): Nutrient value in fresh vs. stored leaves. *J. Food Comp. Anal.* **2016**, *51*, 76–84. [CrossRef]

19. Cefola, M.; D'Antuono, I.; Pace, B.; Calabrese, N.; Carito, A.; Linsalata, V.; Cardinali, A. Biochemical relationships and browning index for assessing the storage suitability of artichoke genotypes. *Food Res. Int.* **2012**, *48*, 397–403. [CrossRef]

20. Kim, J.G.; Luo, Y.; Tao, Y.; Saftner, R.A.; Gross, K.C. Effect of initial oxygen concentration and film oxygen transmission rate on the quality of fresh-cut romaine lettuce. *J. Sci. Food Agric.* **2005**, *85*, 1622–1630. [CrossRef]

21. Fadda, A.; Pace, B.; Angioni, A.; Barberis, A.; Cefola, M. Suitability for ready-to-eat processing and preservation of six green and red baby leaves cultivars and evaluation of their antioxidant value during storage and after the expiration date. *J. Food Process. Pres.* **2016**, *40*, 550–558. [CrossRef]

22. Degl'Innocenti, E.; Pardossi, A.; Tognoni, F.; Guidi, L. Physiological basis of sensitivity to enzymatic browning in "lettuce", "escarole" and "rocket salad" when stored as fresh-cut products. *Food Chem.* **2007**, *104*, 209–215. [CrossRef]

23. Zhou, T.; Harrison, A.D.; McKellar, R.; Young, J.C.; Odumeru, J.; Piyasena, P.; Lu, X.; Mercer, D.G.; Karr, S. Determination of acceptability and shelf life of ready-to-use lettuce by digital image analysis. *Food Res. Int.* **2004**, *37*, 875–881. [CrossRef]

24. Pace, B.; Cefola, M.; Da Pelo, P.; Renna, F.; Attolico, G. Nondestructive evaluation of quality and ammonia content in whole and fresh-cut lettuce by computer vision system. *Food Res. Int.* **2014**, *64*, 647–655. [CrossRef]

25. Martínez-Sánchez, A.; Tudela, J.A.; Luna, C.; Allende, A.; Gil, M.I. Low oxygen levels and light exposure affect quality of fresh-cut Romaine lettuce. *Postharvest Biol. Technol.* **2011**, *59*, 34–42. [CrossRef]

26. Cantwell, M.; Suslow, T. Lettuce: Crisphead or Iceberg. Recommendations for Maintaining Postharvest Quality. 2001. Available online: http://ucanr.edu/sites/Postharvest_Technology_Center_/Commodity_Resources/Fact_Sheets/Datastores/Vegetables_English/?uid=19&ds=799 (accessed on 21 May 2020).

27. Lopez-Gàlvez, G.; Salveit, M.; Cantwell, M. The visual quality of minimally processed lettuces stored in air or controlled atmosphere with emphasis on romaine and iceberg types. *Postharvest Biol. Technol.* **1996**, *8*, 179–190. [CrossRef]

28. Gorny, J.R. A summary of CA and MA requirements and recommendations for fresh-cut (minimally processed) fruits and vegetables. In Proceedings of the 7th International Controlled Atmosphere Conference, University of California, Davis, CA, USA, 13–18 July 1997; Volume 5, pp. 30–66.

29. Hertog, M.L.A.T.; Peppelenbos, H.W.; Evelo, R.G.; Tijskens, L.M.M. A dynamic and generic model of gas exchange of respiring produce: The effects of oxygen, carbon dioxide and temperature. *Postharvest Biol. Technol.* **1998**, *14*, 335–349. [CrossRef]

30. Ruíz-Jiménez, J.M.; Zapata, P.J.; Serrano, M.; Valero, D.; Martínez-Romero, D.; Castillo, S.; Guillén, F. Effect of oxalic acid on quality attributes of artichokes stored at ambient temperature. *Postharvest Biol. Technol.* **2014**, *95*, 60–63. [CrossRef]

31. Ali, H.M.; El-Gizawy, A.M.; El-Bassiouny, R.E.; Saleh, M.A. The role of various amino acids in enzymatic browning process in potato tubers, and identifying the browning products. *Food Chem.* **2016**, *192*, 879–885. [CrossRef]

© 2020 by the authors. Licensee MDPI, Basel, Switzerland. This article is an open access article distributed under the terms and conditions of the Creative Commons Attribution (CC BY) license (http://creativecommons.org/licenses/by/4.0/).

 *foods*

Article

Effect of *Kelulut* Honey Nanoparticles Coating on the Changes of Respiration Rate, Ascorbic Acid, and Total Phenolic Content of Papaya (*Carica papaya* L.) during Cold Storage

Bernard Maringgal [1,2], Norhashila Hashim [1,3,*], Intan Syafinaz Mohamed Amin Tawakkal [4], Mahmud Tengku Muda Mohamed [5], Muhammad Hazwan Hamzah [1] and Maimunah Mohd Ali [1]

1 Department of Biological and Agricultural Engineering, Faculty of Engineering, Universiti Putra Malaysia, 43400 Serdang, Selangor, Malaysia; bernardmaringgal@yahoo.com (B.M.); hazwanhamzah@upm.edu.my (M.H.H.); maimunah_mohdali@ymail.com (M.M.A.)
2 Department of Agriculture Malaysia, Putrajaya 62624, Malaysia
3 SMART Farming Technology Research Centre (SFTRC), Faculty of Engineering, Universiti Putra Malaysia, Serdang 43400, Selangor, Malaysia
4 Department of Process and Food Engineering, Faculty of Engineering, Universiti Putra Malaysia, Serdang 43400, Selangor, Malaysia; intanamin@upm.edu.my
5 Department of Crop Science, Faculty of Agriculture, Universiti Putra Malaysia, Serdang 43400, Selangor, Malaysia; mtmm@upm.edu.my
* Correspondence: norhashila@upm.edu.my

Citation: Maringgal, B.; Hashim, N.; Mohamed Amin Tawakkal, I.S.; Mohamed, M.T.M.; Hamzah, M.H.; Mohd Ali, M. Effect of *Kelulut* Honey Nanoparticles Coating on the Changes of Respiration Rate, Ascorbic Acid, and Total Phenolic Content of Papaya (*Carica papaya* L.) during Cold Storage. *Foods* **2021**, *10*, 432. https://doi.org/10.3390/foods10020432

Academic Editor: Lorenzo Zacarías, Bernardo Pace and Maria Cefola
Received: 22 November 2020
Accepted: 19 January 2021
Published: 16 February 2021

Publisher's Note: MDPI stays neutral with regard to jurisdictional claims in published maps and institutional affiliations.

Copyright: © 2021 by the authors. Licensee MDPI, Basel, Switzerland. This article is an open access article distributed under the terms and conditions of the Creative Commons Attribution (CC BY) license (https://creativecommons.org/licenses/by/4.0/).

Abstract: This study evaluated the respiration rate of coated and uncoated (control) papayas (*Carica papaya* L.) with 15% of *Kelulut* honey (KH) nanoparticles (Nps) coating solution during cold storage at 12 ± 1 °C for 21 days. The respiration rate of the papayas significantly changed during storage, with an increase in CO_2 and a decrease in O_2 and C_2H_4, while the ascorbic acid and total phenolic content was maintained. The changes in respiration rate were rather slower for coated papayas when compared to control ones. A kinetic model was established from the experimental data to describe the changes of O_2, CO_2, and C_2H_4 production in papayas throughout the storage period. All O_2, CO_2, and C_2H_4 were experimentally retrieved from a closed system method and then represented by the Peleg model. The outcomes indicated the Peleg constant K_1 and K_2, which were gained from linear regression analysis and coefficients of determination (R^2), seemed to fit well with the experimental data, whereby the R^2 values exceeded 0.85 for both coated and control papayas. The model confirmed both the capability and predictability aspects of the respiration rate displayed by papayas coated with KH Nps throughout the cold storage period. This is supported by the differences in the stomatal aperture of coated and control papaya shown by microstructural images.

Keywords: kinetic model; Peleg constant; papaya; respiration rate; nanoparticles coating; shelf life

1. Introduction

Papaya is a tropical climacteric fruit with high respiration rates and ethylene (C_2H_4) production during ripening. The fruit is rich in vitamins, minerals, and dietary antioxidants [1]. Nevertheless, it has a short life span due to its climacteric respiration pattern. Papaya of colour indices 2 and 3 can last between five and seven days at ambient temperature, while those with colour indices 4 and 5 can maintain their quality for only two to three days [2]. The short life span enhances the rate of natural deterioration, such as physicochemical damages that eventually increase its susceptibility to diseases and infection [3]. This results in post-harvest loss, a deficit in production yield, as well as a limitation to long-distance export destinations.

The two post-harvest handling strategies, namely modified atmosphere packaging (MAP) and controlled atmosphere (CA), can effectively control the quality of fresh produce by modifying the gas exchange of the fresh produce and its nearby atmosphere [4–6]. Edible

39

coating, an effective MAP strategy, has displayed exceptional outcomes for maintaining the quality of tropical climacteric fruits such as papaya [5,7], mango [8,9], and banana [10,11]. According to Maringgal et al. [12], an edible coating is capable of creating a semipermeable barrier on fruit surfaces that can decrease the fruit respiration rate and C_2H_4 production and result in the maximum quality retention of fruit during storage. Several studies have reported that edible coating can increase the shelf life of papaya during storage by decreasing respiration rate, slowing down the senescence process, and maintaining the chemical compositions, such as ascorbic acid (AA) and total phenolic content (TPC) of papaya [13–16].

Fruit respiration rate can be described by its carbon dioxide (CO_2) generation and oxygen (O_2) consumption. Recently, Maringgal et al. [17] have suggested that the *Kelulut* honey (KH) coating can potentially minimise the papaya respiration rate during cold storage at 12 ± 1 °C and prolong their shelf life up to 12 days. The authors also added that KH coating can act as a physical barrier that leads to the rapid CO_2 accumulation rate of the coated fruit. On the other hand, Nicolaï et al. [18] explained that CO_2 may have some direct and indirect controlling effect on the respiration metabolism and C_2H_4 production. The respiration rate can be determined via a flow-through system or closed system [19]. For the flow-through system, the fruit sample is placed in a non-permeable container where a gas mixture flows at a constant rate. In the closed system, the fruit sample is filled with gas-tight containers of known volume which contain ambient air as the initial atmosphere.

Respiration rate modelling is an effective method to evaluate the respiratory kinetics of fresh produce. There are several respiration rate models that have been used in fruits, such as the Arrhenius, Michaelis–Menten, and Peleg models [17,20]. The Peleg model has been applied to analyse the experimental data on the changing of O_2, CO_2, and C_2H_4 concentrations as well as describe the rate of gas composition function. The suitability of the Peleg model in describing respiration rate has been acknowledged for some fruits, including apples [21,22], bananas [23,24], fresh-cut papaya [20], mangoes [25], and kiwiberry [26]. However, studies on using Peleg modelling to determine the impact of nano-coating on the rate of respiration for the whole papaya are almost non-existent. Most of the studies have focused more on its potential in enhancing the quality of the fruit [27–29]. For instance, Vieira et al. [30] examined the efficacy of hydroxypropyl methylcellulose coating incorporated with silver nanoparticles (Nps) on the post-harvest shelf life of papaya. It was found that nano-coating successfully maintained the quality of papayas, extended the shelf life of papaya up to 14 days, controlled the development of anthracnose disease, and reduced the rate of respiration of the fruit during storage. According to Maringgal et al. [31], this could be due to the reduced Nps size that leads to superior Nps penetration, absorption, as well as migration into fruits, thus promoting an effective coating system.

Therefore, this study aims to (1) identify the impact of nano-coating on papaya's respiration rate, AA, as well as TPC, and (2) investigate the respiration rate kinetic model of nano-coated papaya using the Peleg model in order to describe the function of gas composition and storage day.

2. Materials and Methods

2.1. Bio-Synthesis of Kelulut Honey Nanoparticles

The KH Nps were generated as described by Maringgal et al. [31] through deposition as well as precipitation of calcium carbonate ($CaCO_3$) in KH via biosynthesis.

2.2. Fruit Sample Preparation

A total of 30 fresh papayas (*Carica papaya* L. cv. Sekaki) at colour index 2 (green with traces of yellow) were employed for this research. The fruit were bought from a commercial wholesaler located in Selangor, Malaysia. The chosen papayas were ensured to be uniform in size and shape, with an average weight ranging between 1000 g and 1500 g. The chosen samples also did not have external injuries and pathogenic infection, and were given additional preventive treatment by immersing into 0.01% chlorinated water that was prepared from 5% sodium hypochlorite. The treatments were divided into two categories, namely control (uncoated) and coated with KH Nps at 15% concentration, with each category consisting of 15 fruit.

Nano-coating of KH Nps at 15% concentration (*w/v*) containing 1% glycerol and Tween 20 were prepared as described by Maringgal et al. [31]. The papayas were dipped into the KH Nps concentrations for 1 min and then air-dried for a duration of 2 h. The samples were packed in commercial corrugated boxes in a single layer and stored at 12 ± 1 °C with 85–90% relative humidity for 21 days. The observed changes in respiration rate, C_2H_4 production, AA, and TPC were recorded at seven-day intervals in which three samples were used for each analysis.

2.3. Determination of Gas Exchange and Ethylene (C_2H_4) Production

The papaya's respiration rate was determined by performing methods detailed by Maringgal et al. [17] with minor modification. A closed system was adopted, whereby the fruit was enclosed for 5 h (previously 2 h) in an airtight container at room temperature. All samples were subjected to the measurement of internal gas concentration (O_2, CO_2, and C_2H_4), whereby 1 mL sample of internal gas from the central cavity of the papaya was drawn out thrice with the use of a syringe after being incubated for 5 h inside the airtight container. A gas chromatography instrument (Agilent 6890N Network, California, CA, USA) equipped with a stainless-steel column (Carboxen-1010 PLOT, 30 m × 0.53 mm I.D) and a thermal conductivity detector (TCD/methanizer-FID, 230 °C) was used, into which the gas samples were injected. The carrier gas, helium, was employed at 3.0 mL·min^{-1} flow rate. Before and after sample analyses, the peak areas that were identified for standard gas mixtures were measured. The measurements of the initial and final O_2, CO_2, and C_2H_4 were carried out and employed to compute the respiration rate. The measurements used the expression of nmol kg^{-1} s^{-1} for O_2 and CO_2, and µL/kg·h for C_2H_4.

2.4. Ascorbic Acid (AA) Assay

The AA was identified according to Wall [32], whereby 20 g of papaya sample was homogenised with 80 mL of 3% metaphosphoric acid (HPO_3) by using a high-speed blender for a minute. The AA was calculated using Equation (1) and expressed in mg/100 g.

$$AA = \frac{\text{mL } dye\ used\ \times\ dye\ factor\ \times\ volume\ of\ product\ (100\ \text{mL})\ \times\ 100}{Weight\ of\ sample\ (20\ \text{g})\ \times\ volume\ of\ sample\ for\ titration\ (5\ \text{mL})} \tag{1}$$

2.5. Determination of the Total Phenolic Content (TPC)

The determination of the TPC was carried out by using the Folin–Ciocalteu reagent prescribed by Abu Bakar et al. [33] and modified by Mendy et al. [3]. First, 300 µL of papaya extract was mixed with 2250 µL of Folin–Ciocalteu reagent, vortexed for 15 s, and allowed to stand for 5 min at room temperature. Next, the mixture was added to 2250 µL of sodium bicarbonate solution (60 g/L). After that, the mixture was allowed to stand for 2 h (previously 90 min) in the dark at room temperature, whereby the absorbance was computed with the use of a Multiskan GO microplate spectrophotometer (Thermo Scientific 1510, Vantaa, Finland) at 750 nm. The outcomes were expressed as mg of gallic acid equivalent per 100 g for the fresh sample (mg GAE/100 g).

2.6. Microstructural Microscopy

For microstructure observation, the papaya peel cubes were fixed in a solution containing 4% glutaraldehyde for 2 days at 4 °C. With the use of 0.1 M sodium cacodylate buffer (pH 7.6), the peel cubes were rinsed three times for 30 min, one after another. In a solution containing 1% (*w/v*) osmium tetraoxide, the papaya peel cubes were post-fixed at 4 °C for 2 h. Using 0.1 M sodium cacodylate buffer solution (pH 7.6), every fixed papaya peel cube was again rinsed three times for 30 min. In a series of acetone concentrations ranging from 35% to 100%, the cubes were dehydrated during three changes of acetone solution. Next, a specimen basket was used to transfer the papaya peel cubes. It was then placed into a critical point dryer (LEICA EM CPD030, Wetzlar, Germany) for 30 min. The drying papaya peel cubes were mounted onto the stub using double-sided adhesive and gold-coated by a gold sputter coater. The samples were observed using a scanning electron microscope (JEOL JSM-IT100, InTouchScope™, Tokyo, Japan).

2.7. Data Analysis, Experimental Data Modelling, and Model Parameter Estimation

The two-way analysis of variance (ANOVA) was applied in determining the statistical variances among the papaya samples and storage days. The mean significant variances between the parameters were compared using the least significant difference (LSD) test at a $p < 0.05$ significance level. Statistical analyses were carried out with the use of SAS 9.4 software (Version 9.4, SAS Institute, Cary, NC, USA).

The experimental data of O_2, CO_2, and C_2H_4 were determined by measuring the gas O_2, CO_2, and C_2H_4 concentrations over the storage days. Changes observed in the evaluated gases over storage days were calculated by employing the closed system equations, as given in the following [23]:

$$R_{O_2} = \frac{\left(y_{O2}^{ti} - y_{O2}^{tf}\right) \times V}{100 \times M \times t} \tag{2}$$

$$R_{CO_2} = \frac{\left(y_{CO2}^{ti} - y_{CO2}^{tf}\right) \times V}{100 \times M \times t} \tag{3}$$

$$R_{C_2H_4} = \frac{\left(y_{C_2H4}^{ti} - y_{C_2H4}^{tf}\right) \times V}{1000 \times M \times t} \tag{4}$$

where R_{O_2} is the O_2 rate, nmol kg^{-1} s^{-1}; R_{CO_2} is the CO_2 rate, nmol kg^{-1} s^{-1}; $R_{C_2H_4}$ is the C_2H_4 rate, µL/kg·h; $\left(y_{O2}^{ti} - y_{O2}^{tf}\right)$ is the O_2 concentration, %; $\left(y_{CO2}^{ti} - y_{CO2}^{tf}\right)$ is the CO_2 concentration, %; $\left(y_{CH4}^{ti} - y_{CH4}^{tf}\right)$ is the C_2H_4 concentration, %; V is the volume of the incubation airtight container, m^3; M is the mass of the fruit, kg; and t is the incubation time, h.

The respiratory quotient (RQ) was determined as the CO_2 respiration rate divided by the O_2 respiration rate. The data were plotted using a response surface analysis (Design Expert Software, Version 11, Stat-Ease Inc., Minneapolis, MN, USA).

3. Results and Discussion

3.1. Changes in Gas Concentrations

Table 1 shows the main and interaction effects of KH Nps coating and storage duration on the gas concentration, respiration rate, RQ, AA, and TPC of papaya fruits at $12 \pm 1\,°C$. On the other hand, Table 2 presents a two-way ANOVA of KH Nps coating and storage durations on the experimental parameters of the papaya fruits. The ANOVA result showed the significant effect of the experimental factors (KH Nps coating and storage durations) on the experimental parameters at $p < 0.05$. Changes in the concentrations of O_2, CO_2, and C_2H_4 during the storage period at $12 \pm 1\,°C$ were observed, as illustrated in Figure 1.

Table 1. Main and interaction effects of *Kelulut* honey (KH) nanoparticles (Nps) coating and storage durations (0, 7, 14, and 21 days) on the gas concentration, respiration rate, respiratory quotient (RQ), ascorbic acid (AA), and total phenolic content (TPC) of papaya fruits at 12 ± 1 °C.

Factors% KH Nps Coating (C)	Gas Concentration			Respiration Rate			RQ	AA (mg/100 g)	TPC (mg GAE/100 g FW)
	O_2 %	CO_2%	$C_2H_4\,\mu L/L$	O_2nmol kg^{-1} s^{-1}	CO_2nmol kg^{-1} s^{-1}	$C_2H_4\,\mu L/kg\cdot h$			
0—Control	23.44 A	0.97 B	17.82 B	51.07×10^{-3} A	2.12×10^{-3} B	14.01 B	0.040 B	53.72 A	26.28 B
15—Coated	21.19 B	1.81 A	12.67 A	42.5×10^{-3} B	3.66×10^{-3} A	9.21 A	0.087 A	49.63 B	28.65 A
Storage duration (days), (SD)									
0	24.83 A	0.68 D	11.47 D	51.13×10^{-3} A	1.40×10^{-3} D	8.54 D	0.03 D	47.95 D	8.82 D
7	22.73 B	1.16 C	13.70 C	47.39×10^{-3} B	2.39×10^{-3} C	10.35 C	0.05 C	49. 96 C	34.95 B
14	21.56 C	1.62 B	16.76 B	45.58×10^{-3} BC	3.37×10^{-3} B	12.81 B	0.08 B	52.96 B	40.32 A
21	20.17 D	2.08 A	19.05 A	43.20×10^{-3} C	4.39×10^{-3} A	14.72 A	0.11 A	55.83 A	25.80 C
C × SD	*	*	*	*	*	*		*	*

* significant at $p < 0.05$. Data represent the mean and three replicates. Different letters are significantly different at $p < 0.05$ by least significant difference (LSD) test. GAE: gallic acid equivalent. FW: fresh weight.

Table 2. Analysis of variance (ANOVA) of KH Nps coating and storage durations on the gas concentration, respiration rate, RQ, AA, and TPC of papaya fruits.

Factor	Parameter	Mean Square	F-Value	Pr < F
Day	O_2 Concentration	23.28	148.71	<0.001
	CO_2 Concentration	2.17	262.79	<0.001
	C_2H_4 Concentration	66.83	62.44	<0.001
	O_2 Rate	0.000069	9.60	0.0007
	CO_2 Rate	0.000009	321.02	<0.001
	C_2H_4 Rate	44.29	44.67	<0.001
	RQ	0.007	256.95	<0.001
	AA	71.43	83.64	<0.001
	TPC	0.022	38.12	<0.001
Coating	O_2 Concentration	30.28	193.44	<0.001
	CO_2 Concentration	4.19	508.46	<0.001
	C_2H_4 Concentration	158.74	148.31	<0.001
	O_2 Rate	0.000468	64.95	<0.001
	CO_2 Rate	0.000014	462.00	<0.001
	C_2H_4 Rate	138.53	139.71	<0.001
	RQ	0.013	448.00	<0.001
	AA	100.49	117.66	<0.001
	TPC	0.003	6.19	0.0218
Day × Coating	O_2 Concentration	20.97	6.98	<0.001
	CO_2 Concentration	0.38	46.24	<0.001
	C_2H_4 Concentration	9.28	8.67	0.0012
	O_2 Rate	0.000022	3.09	0.0569
	CO_2 Rate	0.000002	51.90	<0.001
	C_2H_4 Rate	6.34	6.39	0.0047
	RQ	0.002	72.76	<0.001
	AA	16.84	19.72	<0.001
	TPC	0.002	3.15	0.0368

The changes in gas concentrations of coated papayas were significantly ($p < 0.05$) delayed, when compared to the control ones. A gradual reduction in O_2 concentration was noted for all the treatments throughout the 21 days (see Figure 1A). Nonetheless, escalating trends were observed for CO_2 and C_2H_4 concentrations in both coated and control papayas (see Figure 1B,C). For coated papaya, the initial CO_2 percentage increased from 0.86% to 2.82%, whereas the control papaya recorded an increase in CO_2 from 0.51% to 1.35%. The production of C_2H_4 was higher on day 21, with the value of 22.92 µL/L and 15.19 µL/L for control and coated papayas, respectively. The increased CO_2 concentration for KH Nps-coated papayas was in line with the reduction in O_2 concentration during storage. Xu et al. [34] reported that the accumulation of CO_2 in coated fruits was because of a reduced respiration rate that inhibited physiological effects towards the fruits. Meindrawan et al. [35] revealed that nanocomposite coating can decrease the amount of O_2 for respiration and limit diffusion of CO_2 out of the fruit tissues. The KH Nps coating might have restricted the gas exchange through papaya peel, thus retarding C_2H_4 production. The study outcomes are in agreement with prior observational studies, which reported that high internal CO_2 concentration in fruit can hinder C_2H_4 generation, thus delaying the ripening of fruit [36].

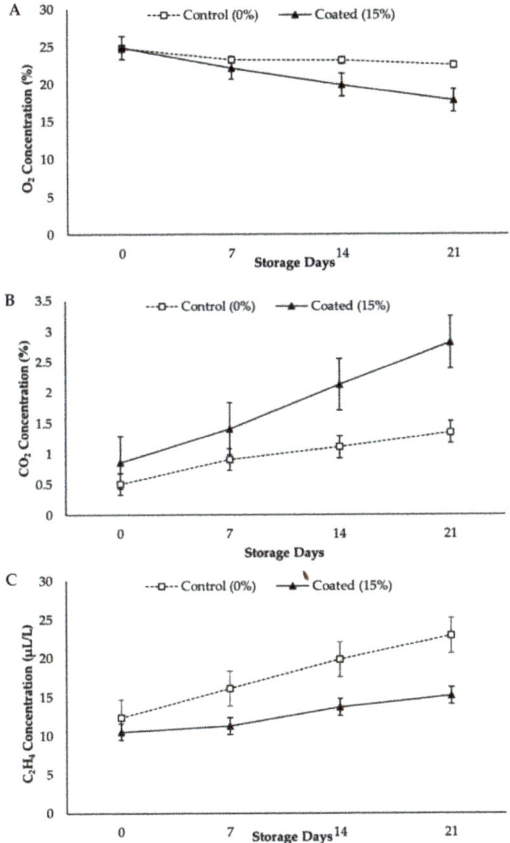

Figure 1. Changes in gas concentration in terms of (**A**) O_2, (**B**) CO_2, and (**C**) C_2H_4 at 0, 7, 14, and 21 days. Vertical bars indicated LSD ($p < 0.05$). LSD: least significant difference.

3.2. Respiration Rate

Both respiration rate and C_2H_4 production rate in fruit are regarded as good indices to determine shelf life [37]. Figure 2 illustrates the significant decrease ($p < 0.05$) in respiration rate in the coated papaya but a gradual increase in respiration rate in the control papaya. This signifies that the O_2 rate in the coated papaya was lower during cold storage when compared to the control samples (see Figure 2A). After 21 days of storage, the highest O_2 rates were 49.78×10^{-3} nmol kg^{-1} s^{-1} and 36.66×10^{-3} nmol kg^{-1} s^{-1} for the control papaya and the coated papaya, respectively.

The effect of KH Nps coating on CO_2 rate appeared to vary significantly ($p < 0.05$). The CO_2 rate was found to increase in parallel with the storage period for both the control and the coated papayas (see Figure 2B). The coated papayas showed the highest CO_2 rate, at 5.79×10^{-3} nmol kg^{-1} s^{-1} on day 21 from its initial rate at 1.71×10^{-3} nmol kg^{-1} s^{-1}. Likewise, the control papayas displayed the lowest CO_2 rate after 21 days of storage, at 2.98×10^{-3} nmol kg^{-1} s^{-1}, whereas the initial CO_2 rate was 1.09×10^{-3} nmol kg^{-1} s^{-1}.

The C_2H_4 production rate differed significantly ($p < 0.05$) due to the KH Nps coating treatment (see Figure 2C). The C_2H_4 production rate in control fruit escalated rapidly and reached its peak after 21 days of storage at 18.21 µL/kg·h. On the contrary, the coated papaya exhibited the lowest value for C_2H_4 production rate after 21 days of storage

(11.24 μL/kg·h) as well as a slow increase in C_2H_4 production rate upon completion of the storage period.

Such delays in respiration and C_2H_4 production rates for the coated papayas, in comparison to those of the control, may suggest that the KH Nps edible coating exerted an obstacle to gaseous exchange. This present study discovered that the KH Nps coating generated a modified atmosphere with high CO_2 and low O_2 in the papaya, which reduced both respiratory and C_2H_4 production rates. Past studies have shown that the reduced rates of respiration and C_2H_4 production in coated papaya resulted in delayed senescence [38]. Similarly, Mendy et al. [3] reported that papaya coated with *Aloe vera* gel coating experienced less weight loss during the storage period when compared to control, which was ascribed to the coatings having provided a semi-permeable layer against gas movement and a resulting decrease in respiration rate. The pattern of respiration and C_2H_4 production rates recorded in this study is in agreement with the findings reported by Ali et al. [13], whereby papaya coating with chitosan suppressed both respiration and C_2H_4 production rates through the modification of the fruit's internal atmosphere.

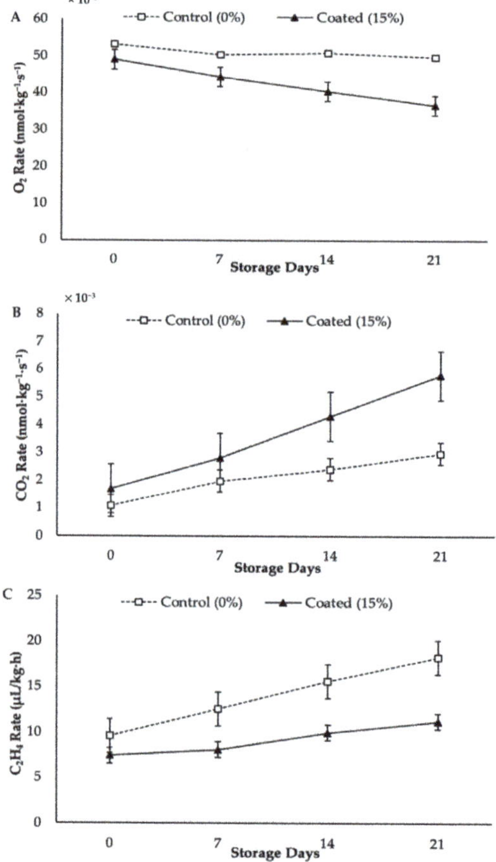

Figure 2. Respiration rate in terms of (**A**) O_2, (**B**) CO_2, (**C**) C_2H_4 production at 0, 7, 14, and 21 days. Vertical bars indicated LSD ($p < 0.05$).

Moreover, Maringgal et al. [17] found that the uncoated papaya displayed an early increase in respiration rate compared to papayas coated with 1.0% and 1.5% KH for 12 days of cold storage period. The reduction in respiration and C_2H_4 production rates due to coating treatment has been reported by many researchers for a range of fruits, including mango [39], mandarin [40], guava [41], and banana [11].

Additionally, Sapper et al. [42] described that the RQ was influenced by changes in respiration rate, indicating the existence of the substrate used during the respiration process of fruit. The RQ was relying on both O_2 and CO_2 production rates, resulting in a decrease in O_2 and an increase in CO_2 [43]. This study revealed the RQ exhibited a significant effect on the storage day and coating treatment of papaya (Tables 1 and 2). The RQ gradually increased for both coated and control papaya during the 21 days of storage, as shown in Figure 3. However, it was noted that the RQ was higher for coated papaya as compared to control papaya. This suggested that the transition of metabolic substrates from carbohydrates to organic acids was higher in coated papaya, which was in agreement with the findings of Sapper et al. [42] and Fagundes et al. [44]. The authors found that an RQ equal to 1 signifies that the metabolic substrates are carbohydrates, while an RQ higher than 1 signifies that the metabolic substrates are organic acids.

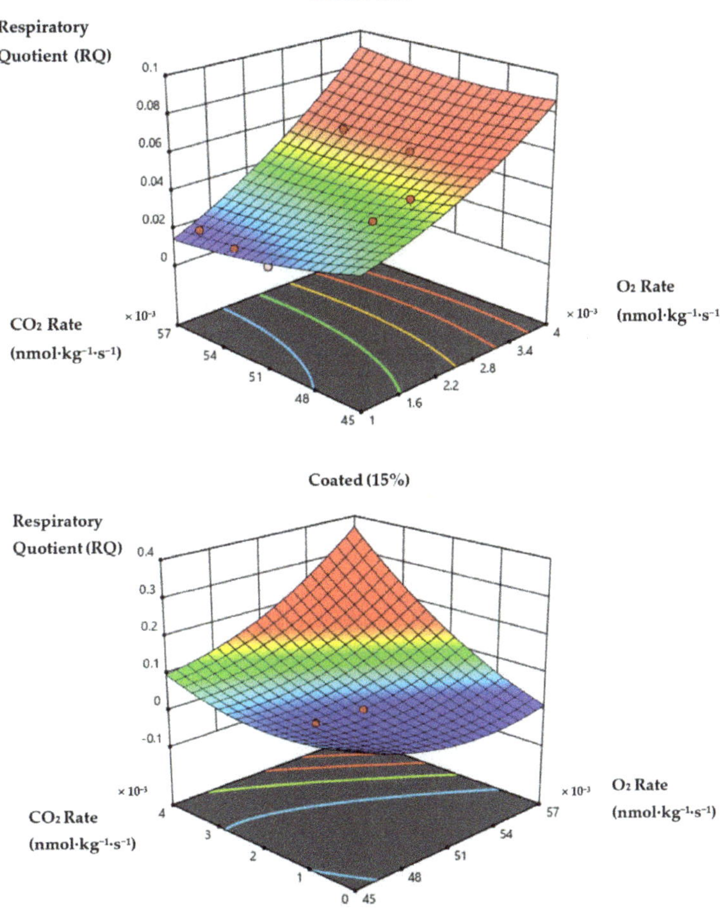

Figure 3. The effect of CO_2 and O_2 partial pressures on the RQ of control and coated papaya fruit at $12 \pm 1\ °C$. The red dot symbols represent data points above and below the response surface, respectively ($p < 0.05$).

3.3. Model Parameter Estimation

The Peleg equations (Equations (2)–(4)), which were modified to reflect curvilinear changes in terms of gas concentration over storage day, were employed to model the computed respiration rates. The inverse of change in gas concentration and the respiration rate models with storage period were plotted, whereby model parameters K_1 and K_2 were estimated via linear regression analysis. The value of K_1 was estimated using the curve's slope, whereas K_2 is denoted by the intercept on the y-axis. Table 3 presents the values of K_1 and K_2, as well as the coefficient of determination (R^2). The R^2 seemed to be very high ($R^2 > 0.85$), which signified an exceptional fit between the experimental data and the Peleg model for coating treatment.

Table 3. Values of Peleg's model parameter (K_1, K_2, and R^2) for O_2, CO_2, and C_2H_4 concentrations in coated and control papaya.

Gas	Papaya	K_1	K_2	R^2
O_2	Control	0.68 ± 0.35	25.15 ± 0.43	0.859
	Coated	2.35 ± 0.32	27.06 ± 0.27	0.997
CO_2	Control	0.27 ± 0.59	0.29 ± 0.61	0.976
	Coated	0.66 ± 0.62	0.16 ± 0.54	0.997
C_2H_4	Control	3.53 ± 0.73	9.01 ± 0.55	0.998
	Coated	1.63 ± 0.77	8.59 ± 0.53	0.967

Both K_1 and K_2 values increased for the model parameters of O_2, which was influenced by the KH Nps coating treatment. For instance, the values of K_1 and K_2 for coated papaya were 2.35 and 27.06, respectively. Comparatively, the values of K_1 and K_2 for control papaya were 0.68 and 25.15, respectively. In this present study, K_2 was more influenced by the coating treatment. A higher value of K_2 was also reported by Rahman et al. [20] for fresh-cut papaya, Bhande et al. [23] for banana, as well as Mahajan and Goswami [21] for apples. The variance between K_1 and K_2 is attributable to both gaseous and physical changes of fruit during the storage period. The Peleg model described K_1 as referring to the rate constant, and it is physically related to both the consumption as well as the evolution of gases starting from the very initial stage, while K_2 refers to the model's capacity constant, implying the respective gas content that can be attained by the system up to infinity. The R^2 values in terms of O_2 were high; 0.997 and 0.859 for coated and control papayas, respectively.

The K_2 appeared to be lower than K_1 for coated papaya in terms of CO_2. The values of K_1 and K_2 for coated papaya were 0.66 and 0.16, respectively; while those for control papaya were 0.27 and 0.29, respectively. The R^2 values in terms of CO_2 were also high; 0.997 and 0.976 for coated and control papayas, respectively.

Next, the relationship between coating treatment and model parameters in terms of C_2H_4 concentration showed a similar trend to the O_2 concentration. For instance, the values of K_1 and K_2 for coated papayas were 1.63 and 8.59, respectively; while those for control papayas were 3.53 and 9.01. The R^2 values in terms of C_2H_4 were 0.967 for coated papayas and 0.998 for control papayas. Figures 4 and 5 illustrate that most of the experimental values were lower than those predicted. The figures display the variances between experimental and predicted respiration rates for coated (see Figure 4) and control (see Figure 5) papayas in terms of O_2, CO_2, and C_2H_4 concentrations. Moreover, the empirical equation developed by Peleg [45] has proven its capability for respiration rate prediction in several fresh produce products [20]. Data developed by this equation plays a significant role in the design of successful modified atmospheric systems for nano-coated papaya. Based on the results of this study, both respiration and C_2H_4 production rates of the papaya were well described by the Peleg kinetic models, and their values have been shown to be significantly influenced by the coating treatments.

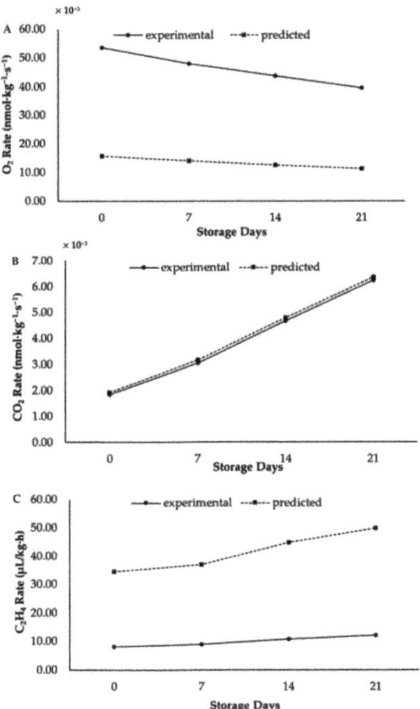

Figure 4. Experimental and predicted respiration rates in term of (**A**) O_2, (**B**) CO_2, and (**C**) C_2H_4 production in KH Nps-coated papaya.

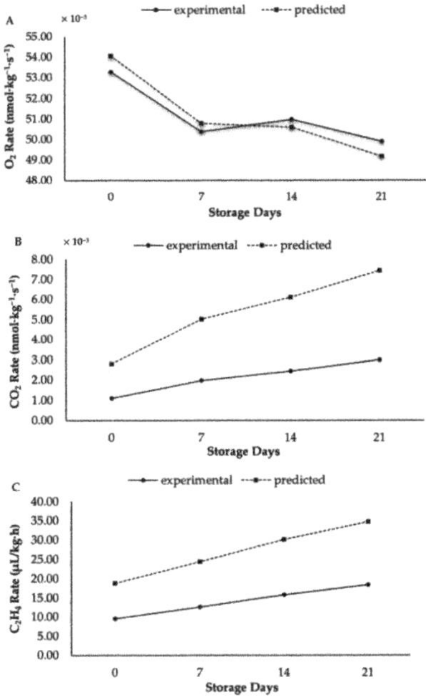

Figure 5. Experimental and predicted respiration rates in term of (**A**) O_2, (**B**) CO_2 and (**C**) C_2H_4 production in control papaya.

3.4. Microscopy Observation

The effect of KH Nps coating on the papaya's surface characteristics was observed by using a scanning electron microscope (SEM). Figure 6A,B indicates the stomatal aperture on papaya peel in control and coated fruits on day 0 of storage. After 21 days of storage, KH Nps coating at 15% concentration continued to cover the stomata on the papaya peel (Figure 6D), hence causing a delay in the ripening process. The KH Nps coating might have protected the stomata on the papaya peels, thus decreasing the respiration rate. Nevertheless, the control papaya peel resulted in exposed stomatal aperture, as shown in Figure 6C, thus reflecting the increase in both respiration and C_2H_4 production rates, apart from affecting both the post-harvest qualities and physicochemical characteristics of papaya throughout the storage period.

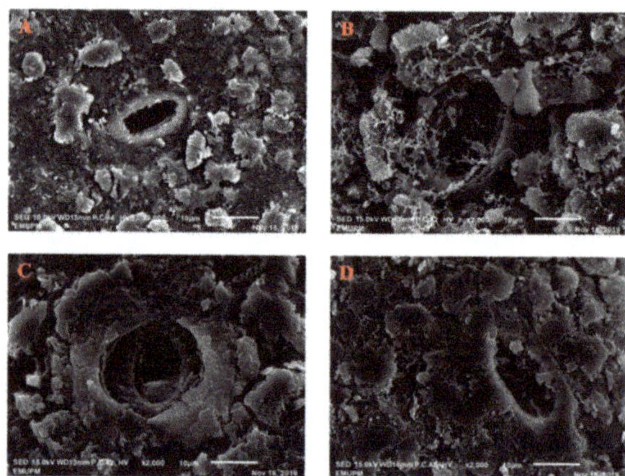

Figure 6. Microstructural images of stomatal aperture on papaya peels at different storage periods: (**A**) control at day 0, (**B**) coated at day 0, (**C**) control at day 21, and (**D**) coated at day 21.

These results were consistent with the research of Jongsri et al. [39], where it was observed that there was a low molecular weight of chitosan coating, which hence conferred protection and covered the stomata of mango peels. This contributes to the delay in the process of ripening during storage. Deng et al. [46] showed that the non-homogeneous coating present on the fruit's surface has the potential to enhance the mass transfer across the fruit surface, thus increasing respiration rate as well as a fungal infection. In the current research, a low respiration rate and delayed C_2H_4 production were maintained by the coated papayas.

3.5. Effect on Ascorbic Acid (AA) and Total Phenolic Content (TPC)

AA, or vitamin C, is a water-soluble vitamin and is an important vitamin in papaya [47,48]. Lee and Kader [49] asserted that AA content can be easily degraded due to improper post-harvest handling and storage conditions. In this study, AA content was significantly ($p < 0.05$) greater for the control papayas than the coated papayas (see Figure 7A). The AA content increased during storage, and the highest AA was recorded for control papaya on day 14 at 59.65 mg/100 g, and 52.01 mg/100 g for coated papaya. This finding is supported by an early study performed by Wills and Widjanarko [50], who explained that AA content will increase in parallel with the ripening stages, then decline thereafter due to senescence.

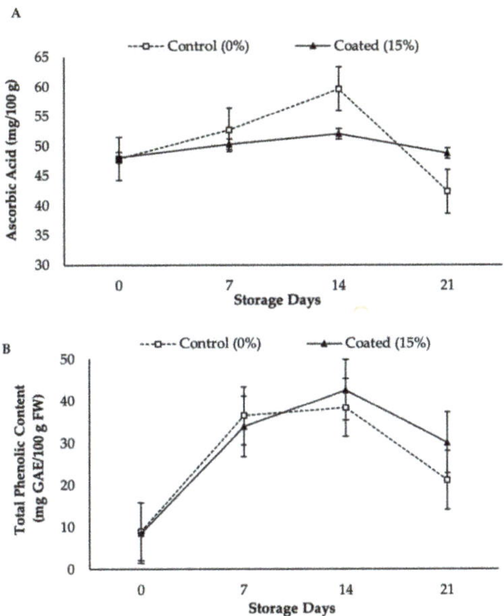

Figure 7. Effect of Ascorbic Acid (**A**) and Total Phenolic Content (**B**) in KH Nps coated papaya stored at $12 \pm 1\,°C$ for 21 storage days. Vertical bars indicated LSD ($p < 0.05$). GAE: gallic acid equivalent. FW: fresh weight.

Furthermore, KH Nps-coated papaya displayed a slower initial increase in AA as compared to the control papaya. At day 21, the AA content of KH Nps-coated papaya maintained a higher value (48.69 mg/100 g) compared to control papaya (42.33 mg/100 g). This suggests that the KH Nps coating slowed down the synthesis of AA during storage. This finding was consistent with Ali et al. [13], wherein the authors found that papaya coating with chitosan is able to reduce the respiration rate of fruit by increasing the CO_2 concentration, thus slowing down the synthesis of AA during storage. The fruit were stored at $12 \pm 1\,°C$ with relative humidity (RH)of 85–90% for four weeks. Additionally, the higher AA retention by KH Nps coating treatment could be associated with the ability of the coating to act as a gas barrier to reduce the O_2 tension in the papaya fruit tissue. This statement was supported by the study of Maftoonazad and Ramaswamy [51], who revealed that the atmosphere composition around the fruit has a significant role in AA retention during storage. Magwaza et al. [47] and Madani et al. [52] added that lower C_2H_4 production in fruits improved AA retention, which was confirmed when C_2H_4 production was measured and revealed a low concentration in KH Nps-coated papaya.

Phenolics are the secondary metabolites existent in plants that have antioxidant properties during the oxidative stress process [53]. It appears that the TPC of papaya varied significantly ($p < 0.05$) due to KH Nps coating (Figure 7B). The highest TPC (42.64 mg GAE/100 g fresh weight (FW)) was recorded for the coated papaya fruits on day 14. The TPC for the control papaya rapidly increased during the storage and reached the maximum (38.48 mg GAE/100 g FW) at day 14, however, it was unable to maintain the TPC rate until day 21 (21.20 mg GAE/100 g FW). A low value of TPC or a sharp reduction in TPC after 14 days in the control papaya could be due to the higher respiration rate, which may have resulted in the loss of TPC because of the degradation of specific phenolic compounds [37].

At the end of storage (day 21), the coated papaya was able to maintain the TPC better than control papaya and this may be because the KH Nps coating reduced the metabolism in the coated papaya. In other words, the KH Nps coating treatment has the ability to maintain the TPC of papaya fruit during cold storage. Similar contributions have been made by Ayón-Reyna et al. [54], who found that untreated papaya had a lower TPC as compared to papaya treated with hydrothermal-calcium chloride. This has also been explored in a prior study by Ghasemnezhad et al. [55], in which the authors revealed the decrease in TPC of coated apricot due to the deterioration process.

4. Conclusions

The kinetic change in respiration rate and the C_2H_4 production of control and coated papayas were studied in this research. The respiration rate and C_2H_4 production in papayas were significantly affected by the KH Nps coating. The current study also showed that KH Nps coating can be used as a conserving material, extending the shelf life by inhibiting the respiration rate and C_2H_4 production, while maintaining the AA and TPC, in papaya. Furthermore, differences in stomatal aperture were observed in coated and control papaya after 21 days of cold storage. This study also suggested that respiration data generated by a closed system could be employed to model the respiration rates. The respiration rate and C_2H_4 production that were predicted by the Peleg model obtained R^2 values higher than 0.85 for both coated and control papaya. This showed that KH Nps is a promising edible coating that can be employed in commercial post-harvest applications for extending the shelf life of papayas.

Author Contributions: Conceptualization, methodology, software, validation, formal analysis, investigation, resources, data curation, writing—original draft preparation, and writing—review and editing: B.M.; supervision, visualization, project administration, funding acquisition, and writing—review and editing: N.H.; supervision and visualization: I.S.M.A.T., M.H.H., and M.T.M.M.; software and validation: M.M.A. All authors have read and agreed to the published version of the manuscript.

Funding: This research was funded by Universiti Putra Malaysia, grant number GP-IPS/9633300 and the article processing charges (APCs) were funded by Universiti Putra Malaysia.

Acknowledgments: The authors express their gratefulness to Universiti Putra Malaysia for the provision of financial support via the Grant Putra research funding (GP-IPS/9633300) that was awarded for this project.

Conflicts of Interest: The authors declare no conflict of interest.

References

1. Singh, S.P.; Sudhakar Rao, D.V. *Papaya (Carica papaya L.)*; Woodhead Publishing Limited: Cambridge, UK; Sawston, UK, 2011; Volume 4, ISBN 9780857090904.
2. MOA Ministry of Agricultural and Agro-Based Industry. 2002. Available online: https://www.moa.gov.my/documents/20182/68591/Pelan+pemasaran+komoditi+betik.pdf/2db80b1b-04a4-42a0-ac53-02cb23345e2b (accessed on 16 November 2020).
3. Mendy, T.K.; Misran, A.; Mahmud, T.M.M.; Ismail, S.I. Application of Aloe vera coating delays ripening and extend the shelf life of papaya fruit. *Sci. Hortic.* **2019**, *246*, 769–776. [CrossRef]
4. Ho, P.L.; Tran, D.T.; Hertog, M.L.A.T.M.; Nicolaï, B.M. Modelling respiration rate of dragon fruit as a function of gas composition and temperature. *Sci. Hortic.* **2020**, *263*, 109138. [CrossRef]
5. Escamilla-García, M.; Rodríguez-Hernández, M.J.; Hernández-Hernández, H.M.; Delgado-Sánchez, L.F.; García-Almendárez, B.E.; Amaro-Reyes, A.; Regalado-González, C. Effect of an edible coating based on chitosan and oxidized starch on shelf life of Carica papaya L., and its physicochemical and antimicrobial properties. *Coatings* **2018**, *8*, 318. [CrossRef]
6. Adiletta, G.; Di Matteo, M.; Albanese, D.; Farina, V.; Cinquanta, L.; Corona, O.; Magri, A.; Petriccione, M. Changes in physico-chemical traits and enzymes oxidative system during cold storage of "Formosa" papaya fresh cut fruits grown in the mediterranean area (Sicily). *Ital. J. Food Sci.* **2020**, *32*, 845–857.
7. Islam, M.Z.; Saha, T.; Monalisa, K.; Hoque, M.M. Effect of starch edible coating on drying characteristics and antioxidant properties of papaya. *J. Food Meas. Charact.* **2019**, *13*, 2951–2960. [CrossRef]
8. Nawab, A.; Alam, F.; Hasnain, A. Mango kernel starch as a novel edible coating for enhancing shelf- life of tomato (*Solanum lycopersicum*) fruit. *Int. J. Biol. Macromol.* **2017**, *103*, 581–586. [CrossRef] [PubMed]
9. Zahedi, S.M.; Hosseini, M.S.; Karimi, M.; Ebrahimzadeh, A. Effects of postharvest polyamine application and edible coating on maintaining quality of mango (*Mangifera indica* L.) cv. Langra during cold storage. *Food Sci. Nutr.* **2019**, *7*, 433–441. [CrossRef]
10. Mendoza, R.; Castellanos, D.A.; García, J.C.; Vargas, J.C.; Herrera, A.O. Ethylene production, respiration and gas exchange modelling in modified atmosphere packaging for banana fruits. *Int. J. Food Sci. Technol.* **2016**, *51*, 777–788. [CrossRef]
11. Thakur, R.; Pristijono, P.; Bowyer, M.; Singh, S.P.; Scarlett, C.J.; Stathopoulos, C.E.; Vuong, Q.V. A starch edible surface coating delays banana fruit ripening. *LWT Food Sci. Technol.* **2019**, *100*, 341–347. [CrossRef]

12. Maringgal, B.; Hashim, N.; Mohamed Amin Tawakkal, I.S.; Muda Mohamed, M.T. Recent advance in edible coating and its effect on fresh/fresh-cut fruits quality. *Trends Food Sci. Technol.* **2020**, *96*, 253–267. [CrossRef]
13. Ali, A.; Muhammad, M.T.M.; Sijam, K.; Siddiqui, Y. Effect of chitosan coatings on the physicochemical characteristics of Eksotika II papaya (*Carica papaya* L.) fruit during cold storage. *Food Chem.* **2011**, *124*, 620–626. [CrossRef]
14. Tabassum, N.; Khan, M.A. Modified atmosphere packaging of fresh-cut papaya using alginate based edible coating: Quality evaluation and shelf life study. *Sci. Hortic.* **2020**, *259*, 108853. [CrossRef]
15. Parven, A.; Sarker, M.R.; Megharaj, M.; Meftaul, I.M. Prolonging the shelf life of Papaya (*Carica papaya* L.) using Aloe vera gel at ambient temperature. *Sci. Hortic.* **2020**, *265*, 109228. [CrossRef]
16. Farina, V.; Passafiume, R.; Tinebra, I.; Scuderi, D.; Saletta, F.; Gugliuzza, G.; Gallotta, A.; Sortino, G. Postharvest application of aloe vera gel-based edible coating to improve the quality and storage stability of fresh-cut papaya. *J. Food Qual.* **2020**, *8303140*, 1–10. [CrossRef]
17. Maringgal, B.; Hashim, N.; Tawakkal, I.S.M.A.; Mohamed, M.T.M.; Hamzah, M.H.; Ali, M.M.; Razak, M.F.H.A. Kinetics of quality changes in papayas (*Carica papaya* L.) coated with Malaysian stingless bee honey. *Sci. Hortic.* **2020**, *267*, 1–11. [CrossRef]
18. Nicolaï, B.M.; Hertog, M.L.A.T.M.; Ho, Q.T.; Verlinden, B.E.; Verboven, P. Gas exchange modeling. In *Modified and Controlled Atmosphere for Storage, Transportation, and Packaging of Horticultural Commodities*; Yahia, E.M., Ed.; CRC Press Taylor & Francis Group: Boca Raton, FL, USA, 2009; pp. 93–110.
19. Fonseca, S.C.; Oliveira, F.A.R.; Brecht, J.K. Modelling respiration rate of fresh fruits and vegetables for modified atmosphere packages: A review. *J. Food Eng.* **2002**, *52*, 99–119. [CrossRef]
20. Rahman, E.A.A.; Talib, R.A.; Aziz, M.G.; Yusof, Y.A. Modelling the effect of temperature on respiration rate of fresh cut papaya (*Carica papaya* L.) fruits. *Food Sci. Biotechnol.* **2013**, *22*, 1581–1588. [CrossRef]
21. Mahajan, P.V.; Goswami, T.K. Enzyme kinetics based modelling of respiration rate for apple. *J. Agric. Eng. Res.* **2001**, *79*, 399–406. [CrossRef]
22. Tappi, S.; Mauro, M.A.; Tylewicz, U.; Dellarosa, N.; Dalla Rosa, M.; Rocculi, P. Effects of calcium lactate and ascorbic acid on osmotic dehydration kinetics and metabolic profile of apples. *Food Bioprod. Process.* **2017**, *103*, 1–9. [CrossRef]
23. Bhande, S.D.; Ravindra, M.R.; Goswami, T.K. Respiration rate of banana fruit under aerobic conditions at different storage temperatures. *J. Food Eng.* **2008**, *87*, 116–123. [CrossRef]
24. Chaguri, L.; Sanchez, M.S.; Flammia, V.P.; Tadini, C.C. Green Banana (*Musa cavendishii*) Osmotic Dehydration by Non-Caloric Solutions: Modeling, Physical-Chemical Properties, Color, and Texture. *Food Bioprocess. Technol.* **2017**, *10*, 615–629. [CrossRef]
25. Ravindra, M.R.; Goswami, T.K. Modelling the respiration rate of green mature mango under aerobic conditions. *Biosyst. Eng.* **2008**, *99*, 239–248. [CrossRef]
26. Bialik, M.; Wiktor, A.; Latocha, P.; Gondek, E. Mass transfer in osmotic dehydration of kiwiberry: Experimental and mathematical modelling studies. *Molecules* **2018**, *23*, 1236. [CrossRef] [PubMed]
27. Castelo Branco Melo, N.F.; de MendonçaSoares, B.L.; Marques Diniz, K.; Ferreira Leal, C.; Canto, D.; Flores, M.A.P.; Henrique da Costa Tavares-Filho, J.; Galembeck, A.; Montenegro Stamford, T.L.; Montenegro Stamford-Arnaud, T.; et al. Effects of fungal chitosan nanoparticles as eco-friendly edible coatings on the quality of postharvest table grapes. *Postharvest Biol. Technol.* **2018**, *139*, 56–66. [CrossRef]
28. Divya, M.; Vaseeharan, B.; Abinaya, M.; Vijayakumar, S.; Govindarajan, M.; Alharbi, N.S.; Kadaikunnan, S.; Khaled, J.M.; Benelli, G. Biopolymer gelatin-coated zinc oxide nanoparticles showed high antibacterial, antibiofilm and anti-angiogenic activity. *J. Photochem. Photobiol. B Biol.* **2018**, *178*, 211–218. [CrossRef]
29. Dhital, R.; Mora, N.B.; Watson, D.G.; Kohli, P.; Choudhary, R. Efficacy of limonene nano coatings on post-harvest shelf life of strawberries. *LWT Food Sci. Technol.* **2018**, *97*, 124–134. [CrossRef]
30. Vieira, A.C.F.; de Matos Fonseca, J.; Menezes, N.M.C.; Monteiro, A.R.; Valencia, G.A. Active coatings based on hydroxypropyl methylcellulose and silver nanoparticles to extend the papaya (*Carica papaya* L.) shelf life. *Int. J. Biol. Macromol.* **2020**, *164*, 489–498. [CrossRef]
31. Maringgal, B.; Hashim, N.; Tawakkal, I.S.M.A.; Hamzah, M.H.; Mohamed, M.T.M. Biosynthesis of CaO nanoparticles using Trigona sp. Honey: Physicochemical characterization, antifungal activity, and cytotoxicity properties. *J. Mater. Res. Technol.* **2020**, *9*, 11756–11768. [CrossRef]
32. Wall, M.M. Ascorbic acid, vitamin A, and mineral composition of banana (*Musa* sp.) and papaya (*Carica papaya*) cultivars grown in Hawaii. *J. Food Compos. Anal.* **2006**, *19*, 434–445. [CrossRef]
33. Abu Bakar, M.F.; Mohamed, M.; Rahmat, A.; Fry, J. Phytochemicals and antioxidant activity of different parts of bambangan (Mangifera pajang) and tarap (*Artocarpus odoratissimus*). *Food Chem.* **2009**, *113*, 479–483. [CrossRef]
34. Xu, D.; Qin, H.; Ren, D. Prolonged preservation of tangerine fruits using chitosan/montmorillonite composite coating. *Postharvest Biol. Technol.* **2018**, *143*, 50–57. [CrossRef]
35. Meindrawan, B.; Suyatma, N.E.; Wardana, A.A.; Pamela, V.Y. Nanocomposite coating based on carrageenan and ZnO nanoparticles to maintain the storage quality of mango. *Food Packag. Shelf Life* **2018**, *18*, 140–146. [CrossRef]
36. Moalemiyan, M.; Ramaswamy, H.S.; Maftoonazad, N. Pectin-based edible coating for shelf-life extension of Ataulfo mango. *J. Food Process. Eng.* **2012**, *35*, 572–600. [CrossRef]
37. Ali, A.; Maqbool, M.; Alderson, P.G.; Zahid, N. Effect of gum arabic as an edible coating on antioxidant capacity of tomato (*Solanum lycopersicum* L.) fruit during storage. *Postharvest Biol. Technol.* **2013**, *76*, 119–124. [CrossRef]

38. Hu, H.; Zhou, H.; Li, P. Lacquer wax coating improves the sensory and quality attributes of kiwifruit during ambient storage. *Sci. Hortic.* **2019**, *244*, 31–41. [CrossRef]
39. Jongsri, P.; Wangsomboondee, T.; Rojsitthisak, P.; Seraypheap, K. Effect of molecular weights of chitosan coating on postharvest quality and physicochemical characteristics of mango fruit. *LWT Food Sci. Technol.* **2016**, *73*, 28–36. [CrossRef]
40. Chen, C.; Peng, X.; Zeng, R.; Chen, M.; Wan, C.; Chen, J. Ficus hirta fruits extract incorporated into an alginate-based edible coating for Nanfeng mandarin preservation. *Sci. Hortic.* **2016**, *202*, 41–48. [CrossRef]
41. Santos, T.M.; Souza Filho, M.d.S.M.; Silva, E.d.O.; Silveira, M.R.S.d.; Miranda, M.R.A.d.; Lopes, M.M.A.; Azeredo, H.M.C. Enhancing storage stability of guava with tannic acid-crosslinked zein coatings. *Food Chem.* **2018**, *257*, 252–258. [CrossRef]
42. Sapper, M.; Palou, L.; Pérez-Gago, M.B.; Chiralt, A. Antifungal Starch-Gellan Edible coatings with thyme essential oil for the postharvest preservation of apple and persimmon. *Coatings* **2019**, *9*, 333. [CrossRef]
43. Beaudry, R.M. Effect of carbon dioxide partial pressure on blueberry fruit respiration and respiratory quotient. *Postharvest Biol. Technol.* **1993**, *3*, 249–258. [CrossRef]
44. Fagundes, C.; Carciofi, B.A.M.; Monteiro, A.R. Estimate of respiration rate and physicochemical changes of fresh-cut apples stored under different temperatures. *Food Sci. Technol.* **2013**, *33*, 60–67. [CrossRef]
45. Peleg, M. An Empirical Model for the Prediction. *J. Food Sci.* **1988**, *53*, 1216–1217. [CrossRef]
46. Deng, Z.; Jung, J.; Simonsen, J.; Zhao, Y. Cellulose nanomaterials emulsion coatings for controlling physiological activity, modifying surface morphology, and enhancing storability of postharvest bananas (*Musa acuminate*). *Food Chem.* **2017**, *232*, 359–368. [CrossRef]
47. Magwaza, L.S.; Mditshwa, A.; Tesfay, S.Z.; Opara, U.L. An overview of preharvest factors affecting vitamin C content of citrus fruit. *Sci. Hortic.* **2017**, *216*, 12–21. [CrossRef]
48. Kathiresan, S.; Lasekan, O. Effects of glycerol and stearic acid on the performance of chickpea starch-based coatings applied to fresh-cut papaya. *CYTA J. Food* **2019**, *17*, 365–374. [CrossRef]
49. Lee, S.K.; Kader, A.A. Preharvest and postharvest factors influencing vitamin C content of horticultural crops. *Postharvest Biol. Technol.* **2000**, *20*, 207–220. [CrossRef]
50. Wills, R.B.H.; Widjanarko, S.B. Changes in physiology, composition and sensory characteristics of australian papaya during ripening. *Aust. J. Exp. Agric.* **1995**, *35*, 1173. [CrossRef]
51. Maftoonazad, N.; Ramaswamy, H.S. Application and Evaluation of a Pectin-Based Edible Coating Process for Quality Change Kinetics and Shelf-Life Extension of Lime Fruit (*Citrus aurantifolium*). *Coatings* **2019**, *9*, 285. [CrossRef]
52. Madani, B.; Mirshekari, A.; Sofo, A.; Tengku Muda Mohamed, M. Preharvest calcium applications improve postharvest quality of papaya fruits (*Carica papaya* L. cv. Eksotika II). *J. Plant. Nutr.* **2016**, *39*, 1483–1492. [CrossRef]
53. Shahidi, F.; Yeo, J.D. Bioactivities of phenolics by focusing on suppression of chronic diseases: A review. *Int. J. Mol. Sci.* **2018**, *19*, 1573. [CrossRef]
54. Ayón-Reyna, L.E.; González-Robles, A.; Rendón-Maldonado, J.G.; Báez-Flores, M.E.; López-López, M.E.; Vega-García, M.O. Application of a hydrothermal-calcium chloride treatment to inhibit postharvest anthracnose development in papaya. *Postharvest Biol. Technol.* **2017**, *124*, 85–90. [CrossRef]
55. Ghasemnezhad, M.; Shiri, M.A.; Sanavi, M. Effect of chitosan coatings on some quality indices of apricot (*Prunus armeniaca* L.) during cold storage. *Casp. J. Environ. Sci.* **2010**, *8*, 25–33.

 *foods*

Article

Effect of Cutting Styles on Quality and Antioxidant Activity of Stored Fresh-Cut Sweet Potato (*Ipomoea batatas L.*) Cultivars

Atigan Komlan Dovene, Li Wang, Syed Umar Farooq Bokhary, Miilion Paulos Madebo, Yonghua Zheng and Peng Jin *

College of Food Science and Technology, Nanjing Agricultural University, Nanjing 210095, Jiangsu, China; 2017108147@njau.edu.cn (A.K.D.); 2017208028@njau.edu.cn (L.W.); 2017208031@njau.edu.cn (S.U.F.B.); 2018208034@njau.edu.cn (M.P.M.); zhengyh@njau.edu.cn (Y.Z.)
* Correspondence: pjin@njau.edu.cn; Tel.: +86-25-84395315; Fax: +86-25-84395618

Received: 14 November 2019; Accepted: 9 December 2019; Published: 12 December 2019

Abstract: The effect of cutting styles (slice, pie, and shred) on the quality characteristics and antioxidant activity of purple and yellow flesh sweet potato cultivars during six days of storage at 4 °C was investigated. The results indicated that the sliced and pie samples showed no significant difference ($p > 0.05$) on the firmness, weight loss, and vitamin C content compared with the whole sweet potato in both cultivars during storage. The pie sample exhibited the highest wound-induced phenolic, flavonoid, and carotenoid accumulation and DPPH radical scavenging activity among the cuts in both cultivars. Moreover, the shredded sample showed significantly ($p < 0.05$) higher polyphenol oxidase (PPO) activity but lower total phenolic and flavonoid content and the lowest antioxidant activity among the samples. Thus, the finding of this study revealed that pie-cut processing has potential in improving the quality and increasing the antioxidant activity of fresh-cut purple and yellow flesh sweet potato cultivars while shredding accelerated the quality deterioration of both sweet potato cultivars.

Keywords: sweet potatoes; cutting styles; quality; antioxidant activity

1. Introduction

Sweet potato (*Ipomoea batatas L.*) root is an important tropical root with significant economic value due to its high nutritional and antioxidant potential [1]. Sweet potato storage root is rich in carbohydrates, dietary fiber, vitamins (A, B1, B2, C, and E) and minerals (Ca, Mg, K, and Zn) [2]. Wounding stress generated by the cut in the plant tissues has been verified to induce the production of phenolic compounds that possess antioxidant activities in plants [3]. This has made cutting an innovative and straightforward technique to increase the accumulation of phenolic compounds and improve the antioxidant activity of many fruits and vegetables during short-term storage [4]. Cutting practices have been explored in carrot [5], pitaya [4], and many other fruits and vegetables. However, the increase in the antioxidant activity of the fresh-cut produce depends on the balance between antioxidants synthesis and oxidation, such as phenolic compounds [6]. This is because the wounding stress triggers two types of responses in phenolic metabolism. Firstly, the oxidation of the existing phenolic compounds due to the disruption of the cell membrane, causing the phenolics to combine with the oxidative enzyme systems, particularly the polyphenol oxidase (PPO) enzyme that is involved in tissue browning. Secondly, the synthesis of monomeric or polymeric phenolics to repair the wounding damage through the phenylpropanoid pathway [5,7,8]. Previous research demonstrated that wounding intensity had an obvious effect on the levels of physiological and biochemical changes in fresh-cut product and cutting styles were one of the most important factors that influence the storage quality and the preservation of fresh-cut fruits and vegetables [9].

According to the reports of Li et al. [4] and Grace et al. [10] the biosynthesis and conversion rates of phenolics and flavonoid in fresh-cut produce changed during storage time, depending on the wounding intensity applied to the plant tissue, hence, it is important to investigate the effect of different cutting styles on the quality and antioxidant activity in fresh-cut fruits and vegetables.

Purple and yellow sweet potato cultivars, in addition to their nutritive components, are also rich in flavonoid and carotenoids pigments that also contribute to its antioxidant capacity [2,11]. The size of sweet potato roots is relatively large; hence fresh-cut sweet potatoes more convenient and preferable for consumers. Fresh-cut sweet potatoes are processed into frozen, dried, canned, fried, fermented, and pureed products and also used as natural food colorants. Currently, there is little information on the effect of cutting styles on the quality characteristics and antioxidant activity of fresh-cut sweet potato cultivars. The purpose of this study is to fill that research gap by evaluating the effect of cutting styles on the quality characteristics and antioxidant activity of fresh-cut sweet potato cultivars.

2. Materials and Methods

2.1. Sample Collection and Processing

Purple and yellow sweet potato cultivars (*Ipomoea batatas* L.) were purchased from the local market in Nanjing, Jiangsu, P.R. China, and transferred to the laboratory in 2 h. The sweet potato cultivars were selected according to uniformity in appearance and absence of physical defects or lesion. The roots from each cultivar were washed, peeled, and cut into three different styles: slice (1 cm thickness), pie or quarter-slice (1/4 section from a slice of 1 cm thickness), and shred. Whole roots were used as control. The samples were then packaged (100 g) in each plastic containers (15 × 10 × 4 cm) and stored at 4 °C with 85–90% relative humidity for six days. During storage, samples (one container of each sample) were taken every two days for firmness, color, and chemical analysis while the weight loss was determined using pre-weighed 500 g of each sample.

2.2. Color Change, Firmness, and Weight Loss Determination

The flesh color of the sweet potato cultivars was determined using a colorimeter (Konica Minolta, Tokyo, Japan) according to the method described by Tang et al. [12]. The color was evaluated by measuring L*, a*, and b* values (L* indicates Lightness, a* red/green coordinate, and b* is the yellow/blue coordinate). The firmness was measured according to the method described by Wall et al. [13] with some modifications. The firmness was measured on two paired sides using a TA-XT2i texture analyzer (Stable Micro System Ltd., London, UK) with a 3 mm diameter probe at a speed of 60 mm/min for slice and pie style while 5 mm diameter probe at a speed of 1 mm/s was used for shred style. Firmness was expressed as newtons (N). The weight was measured in gram using the electronic scale (maximum (max = 300 g), verification scale value (e = 0.1 g), scale division value (d = 0.01 g) (Chengdu Beisaike Instrument Research Institute, Chengdu, China). The weight loss was expressed as a percentage (%). The color was measured from taken from the very same sample-pieces on each sampling

2.3. Total Phenolics and Total Flavonoids Content Determination

The total phenolics contents of the samples were determined using the Folin–Ciocalteau assay method [14]. Absorbance was measured at 765 nm, and the results were expressed as milligrams of gallic acid (GAE) per gram. Total flavonoid contents of the sweet potato cultivars were determined using the vanillin-HCl method [15]. The absorbance of the standard catechin solution was measured at 430 nm, and 80% ethanol was used as control. Total flavonoid content was expressed as catechin equivalent derived from the standard curve.

2.4. Analysis of Total Carotenoids and Vitamin C Contents

Carotenoid content was determined using trichloroacetic acid (TCA) solution extraction followed by spectrophotometric analysis, as described by Huang et al. [16]. Absorbance was measured at 534 nm using a spectrophotometer (TU-1810 DSPC, Beijing Puxi Instrument Co., Beijing, China). Vitamin C content was analyzed by the procedure of Arakawa et al. [17]. The fresh sweet potato (2 g) was extracted in 5 mL of 5% TCA solution. The homogenate extracts were centrifuged at $12,000 \times g$ for 20 min at 4 °C, and then the supernatants were used for vitamin C analysis. Vitamin C content was expressed as mg g^{-1} fresh weight, based on a standard curve.

2.5. Antioxidant Determination

The antioxidant activity of the sweet potato tissue was measured using 2,2-diphenyl-1-picrylhydrazyl (DPPH) free radical-scavenging activity, as reported by Bae et al. [18] with some modifications. The fresh tissue (2 g) was extracted using 5 mL of 50% ethanol. The homogenate was centrifuged at $13,000 \times g$ for 20 min at 4 °C. The enzyme extracts (0.1 mL) was added to 1.9 mL DPPH (120 µmol L^{-1}) and reacted in the dark for 20 min, and the absorbance was measured at 525 nm. The mixture of 0.1 mL of 50% ethanol and 1.9 mL of DPPH reagent was used as the control.

The following equation was used when calculated

$$\% \text{ DPPH scavenging phenol ratio} = 1 - [(A - B)/A_0] \times 100\%$$

A = the absorbance of the sample; B = the absorbance of the sample with 1.9 mL 50% ethanol); and A_0 = the absorbance of the control.

Ferric reducing antioxidant power (FRAP) assay was done according to the method described by Chen et al. [19] with some modification. FRAP solution used for the assay was freshly prepared by mixing 100 mL acetate buffer (0.3 M, pH 3.6) in 10 mL TPTZ solution (10 mM, in 40 mM HCl) and 10 mL ferric chloride (20 mM). This solution was warmed at 37 °C with a water bath before use. Each sample (1 mL) was then added to the freshly prepared FRAP solution (5 mL); the mixture was kept in the dark at 37 °C for 20 min. The absorbance was read at 593 nm against a blank using a microplate spectrophotometer. Different concentrations (100–1400 µM) of ferrous sulfate standard solutions were used to prepare the calibration curve. Higher FRAP value indicated greater ferric reducing antioxidant capacity and final results were expressed as µM Fe (II).

2.6. PPO Activity

The measurement of PPO activity was carried out according to the method described by Manohan et al. [20] with slight modification. The fresh sweet potato tissue (1 g) was ground in 5 mL of phosphate buffer, and the homogenate was centrifuged at $12,000 \times g$ for 20 min at 4 °C. The reaction system consisted of 1.9 mL of phosphate buffer, 1 mL of enzyme solution, and 1 mL of catechol (0.1 mol L^{-1}), and the absorbance was measured at 420 nm.

2.7. Total Aerobic Bacterial Count (TABC)

Total aerobic bacterial count (TABC) was analyzed according to the method by Li et al. [4]. The results were expressed as log10 colony-forming unit (CFU) per kilogram based on fresh weight (log CFU g^{-1}).

2.8. Statistical Analysis

Statistical analyses were performed using SPSS version 20.0 (SPSS INC., Chicago, IL, USA) software data in the research are represented as means ± standard deviation (SD) of three replications. One-way ANOVA analyzed data, and differences among samples were determined by comparison of means using Duncan's multiple range test at $p < 0.05$.

3. Results and Discussion

3.1. Whole, Sliced, Pie, and Shredded Purple and Yellow Flesh Sweet Potato Cultivars

The whole, sliced, pie, and shredded purple and yellow flesh sweet potato cultivars are shown in Figure 1. Interestingly, the different cutting styles induced varying degrees of progressive changes in the color of both sweet potato cultivars.

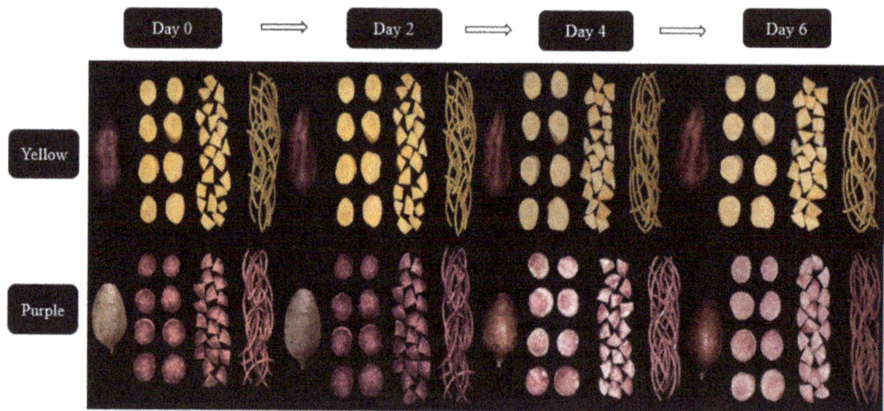

Figure 1. Whole, sliced, pie, and shredded yellow and purple flesh sweet potato cultivars.

The pie-cut had the most significant color changes within the first two days of the storage in both cultivars, but on day 6 of the storage, the shredded samples became the most discolor on visual observation. The higher color changes in the pie samples in both cultivars at the early stage of the storage could be due to increased accumulation of phenolics, flavonoid, and carotenoid. The excessive discoloration, especially of the shredded samples during the late stage of the storage, could be due to wound-induced phenolic oxidation together with increased polyphenol oxidase activity in both sweet potatoes. The purple flesh sweet potato showed more rapid change in color compared to the yellow flesh sweet potato. This could be due to higher wound-induced phenolics, flavonoid, and carotenoid phenolic accumulation in the purple flesh sweet potato compared to the yellow flesh sweet potato. Color and visual appeal is an important quality parameter of fresh-cut produce: while slight changes in color might indicate increased phenolics, carotenoid, flavonoid and/or other phytochemicals accumulation, excessive discoloration is often interpreted as an indication of oxidative deterioration and often affect consumer acceptability of fresh-cut produce [12].

3.2. Effect of Cutting Styles on the Color, Firmness, and Weight Loss of the Sweet Potato Cultivars

Excessive coloration is an indication of low quality in fresh-cut sweet potatoes [21] hence, the determination of color is of great importance in investigating the quality of fresh-cut sweet potato cultivars. The effect of the cutting styles on the color of the sweet potato cultivars is shown in Table 1. Cutting induced the increase in L* value with storage time in both cultivars. The a* value in purple flesh sweet potato was higher than that in yellow flesh sweet potato, and it presented a decreasing trend in purple flesh sweet potato during storage. At the end of storage, pie cutting sweet potatoes showed the lowest a* value among samples in both cultivars. The b* value of purple flesh sweet potato in different cutting styles increased significantly ($p < 0.05$) with storage time, while yellow flesh sweet potatoes exhibited a decreasing trend. The shredded sweet potato cultivars showed the highest b* values at the end of storage. These results indicated that the color of fresh-cut sweet potatoes changed with wounding intensity and storage time (Figure 1). Sweet potato cultivars contained naturally pigmented phytochemicals, including β-carotene, which appears dark green, yellow, or orange-color and flavonoids that display yellow color [10]. The results of total carotenoid and flavonoid contents verified that cutting could induce the synthesis of the pigmented phytochemicals, which may closely relate with the changes in the color of sweet potato cultivars (Figure 2C,D and Figure 3A,B). Besides, the decreasing trends in b* value of slice and pie cutting yellow flesh sweet potato indicated that cutting might also result in the oxidation and degradation of pigmented phytochemicals. Furthermore, the excessive coloration in the shredded samples could also be a result of increased microbial contamination. Microorganisms have been reported to contaminate and discolor store fresh-cut produce [22] The color variations in the purple and yellow flesh sweet potato cultivars are similar to the previously published data on the effect of storage time on the color of sweet potato cultivars [12].

As shown in Table 1, the firmness of both sweet potato cultivar decreased with storage time, and similar results were also found in the research of Wall et al. [13]. Besides, the increased of wounding intensity accelerated the decrease of firmness in sweet potatoes and shred cutting sweet potatoes presented the lowest firmness in both cultivars during storage. The progressive decrease in firmness in the shredded sample could be the result of rupture of the plasma membrane, and plasmolysis in tissue, which indicated the deterioration and loss of quality. Hence fresh-cut produce with extremely low firmness, as seen in the shredded sample in both cultivars, often have low acceptability.

Fresh cut processing resulted in significant ($p < 0.05$) weight loss in the shredded sample of both the purple and yellow flesh sweet potatoes while the sliced and pie samples were not significantly ($p < 0.05$) affected compared to the whole sweet potato. The weight loss in fresh-cut produce was mainly caused by the evaporation moisture and the loss of nutrition. High weight loss in stored sweet potato cultivars indicated short storage-life, rapid deterioration, and loss of quality [23]. In our results, shred cutting accelerated weight loss. The higher weight loss resulted in the lower turgor pressure in the cell, which played an important role in the firmness among samples of different cutting styles. Thus, the results of color, firmness and, weight loss suggested that shred cutting in both cultivars lead to the quality deterioration of the fresh-cut produce while the slice and pie-cut with relatively lower wounding intensity had no adverse on the quality during the storage period.

Table 1. Effect of cutting style on the color, firmness, and weight loss of the purple and yellow flesh sweet potato cultivars.

Parameter		Cutting Style	Purple Flesh Sweet Potato				Yellow Flesh Sweet Potato			
			Storage Time (Day)							
			0	2	4	6	0	2	4	6
Color	L*	Whole	31.68 ± 1.07 a	32.45 ± 2.04 c	32.99 ± 3.01 c	33.53 ± 1.24 b	29.24 ± 1.89 b	67.37 ± 2.48 c	67.82 ± 2.78 c	69.71 ± 1.36 b
		Slice	31.34 ± 1.51 a	48.1 ± 2.43 a	47.77 ± 3.92 b	47.69 ± 1.65 a	30.24 ± 0.89 b	73.37 ± 0.78 a	74.46 ± 0.78 a	74.58 ±0.59 a
		Pie	31.16 ± 1.69 a	45.74 ± 2.08 b	49 ± 1.77 a	46.18 ± 3.56 a	29.8 ± 0.68 b	73.28 ± 0.68 a	75.04 ± 0.46 a	75.21 ±0.82 a
		Shred	30.13 ± 1.24 a	32.95 ± 1.49 c	31.66 ± 1.21 c	32.52 ± 1.66 b	32.03 ± 1.84 a	69.91 ± 2.99 b	69.85 ± 1.65 b	69.93 ±1.68 b
	a*	Whole	17.29 ± 1.05 c	17.8 ±1.5 b	17.41 ± 0.96 a	16.48 ± 1.03 a	8.167 ± 0.72 b	8.8 ± 0.79 c	8.78 ± 0.99 b	8.63 ± 0.52 b
		Slice	19.29 ± 1.4 b	17.4 ± 1.08 b	16.46 ± 0.96 b	13.66 ± 1.58 b	8.37 ± 0.98 b	9.5 ± 0.88 a	9.47 ± 0.99 a	9.37 ± 0.52 a
		Pie	21.72 ± 0.77 a	18.65 ± 0.59 a	17.24 ± 5.14 a	14.22 ± 1.91 b	9.73 ± 0.74 a	9.30 ± 0.91 b	8.98 ± 0.91 b	8.27 ± 0.6 b
		Shred	16.45 ± 0.51 c	17.61 ± 0.52 b	16.88 ± 0.4 b	15.94 ± 0.85 a	9.44 ± 0.94 a	10.14 ± 0.58 a	9.7 ± 0.84 a	9.5 ± 0.88 a
	b*	Whole	1.5 ± 0.28 b	3.98 ± 0.66 b	4.31 ± 0.67 b	4.8 ± 0.69 b	28.72 ± 0.67 a	27.8 ± 0.84 c	26.45 ± 0.89 b	25.3 ± 0.61 a
		Slice	1.4 ± 0.28 b	4.47 ± 0.46 a	4.58 ± 0.47 a	4.7 ± 0.86 b	28.98 ± 0.67 a	25.7 ± 0.44 b	24.67 ± 0.59 c	24.3 ± 0.34 ac
		Pie	1.4 ± 0.67 b	3.8 ± 0.35 b	4.46 ± 0.46 a	4.9 ± 1.09 b	29.89 ± 0.48 a	28.18 ± 0.45 a	25.78 ± 0.52 b	23.84 ± 0.53 c
		Shred	2.37 ± 0.21 a	3.2 ± 0.57 c	3.98 ± 0.27 c	5.22 ± 0.3 a	28.47 ± 0.76 a	28.85 ± 0.42 a	28.79 ± 0.65 a	28.77 ± 0.8 a
Firmness (N)		Whole	198.7 ± 5.67 a	197.16 ± 4.8 a	195.76 ± 2.56 a	183.22 ± 2.91 a	170.06 ± 1.79 a	169.74 ± 2.64 a	168.38 ± 2.05 a	167.05 ± 2.1 a
		Slice	196.3 ± 3.67 a	196.16 ± 1.7 a	195.38 ± 1.56 a	175.52 ± 2.91 b	169.95 ± 0.93 a	169.53 ± 0.98 a	168.45 ± 0.89 a	167.55 ± 0.8 a
		Pie	196.95 ± 1.71 a	196.76 ± 1.48 a	194.9 ± 0.66 a	151.38 ± 2.61 c	169.71 ± 0.95 a	168.48 ± 0.8 a	165.08 ± 0.71 b	159.67 ± 0.9 b
		Shred	196.34 ± 1.84 b	159.91 ± 2.09 b	154.67 ± 2.77 b	134.24 ± 0.75 d	164.62 ± 0.85 b	162.52 ± 0.8 b	159.01± 0.84 c	156.96 ± 0.8 c
Weight loss (%)		Whole	171.17 ± 6.15 a	171.06 ± 4.66 a	170.95 ± 3.08 a	170.15 ± 4.56 a	171.27 ± 1.25 a	171.02 ± 1.63 a	170.01 ± 2.8 a	169.99 ± 4.6 a
		Slice	171.17 ± 2.315 a	170.37 ± 1.56 a	170.11 ± 2.08 a	169.07 ± 3.07 a	171.17 ± 1.25 a	170.64 ± 1.63 a	170.17 ± 2.8 a	169.87 ± 2.0 a
		Pie	171.17 ± 2.25 a	169.67 ± 1.8 a	168.86 ± 2.13 a	168.8 ± 2.15 a	171.57 ± 2.18 a	170.36 ± 3.5 a	169.71 ± 1.6 a	169.39 ± 1.6 a
		Shred	171.17 ± 3.11 a	167.12 ± 1.91 b	165.59 ± 1.45 b	165.46 ± 0.98 b	171.37 ± 1.82 a	169.72 ± 2.11 a	166.89 ± 3.2 b	166.25 ± 2.5 b

Statistical analysis using ANOVA ($n = 3$) at a 95% confidence interval ($p \leq 0.05$) using the Duncan test. The same letter within the same column indicates no significant difference between samples: PFPT—purple flesh sweet potato; YFPT—yellow flesh sweet potato; L*—Lightness, a*—red/green and b*—yellow/blue.

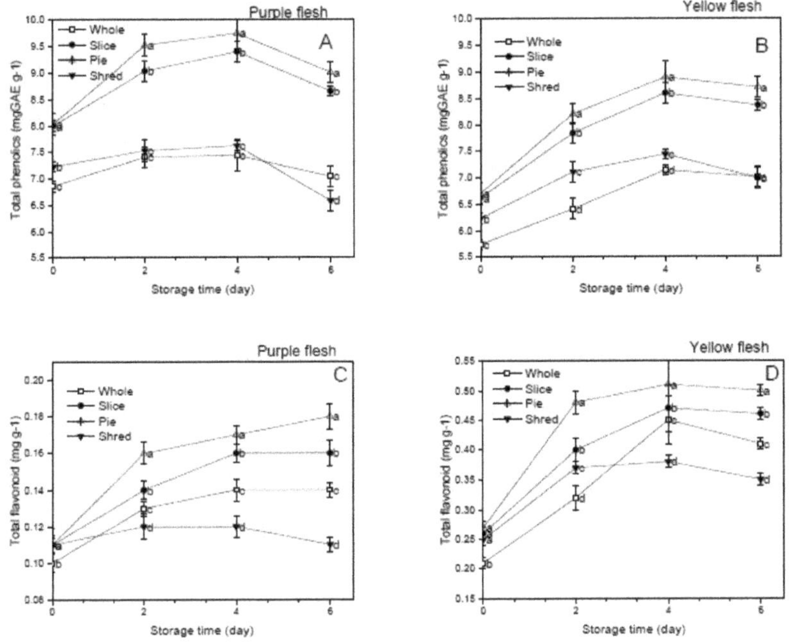

Figure 2. Effect of cutting styles on the total phenolic (**A,B**), total flavonoid (**C,D**) content of the purple and yellow flesh the sweet potato cultivars, respectively.

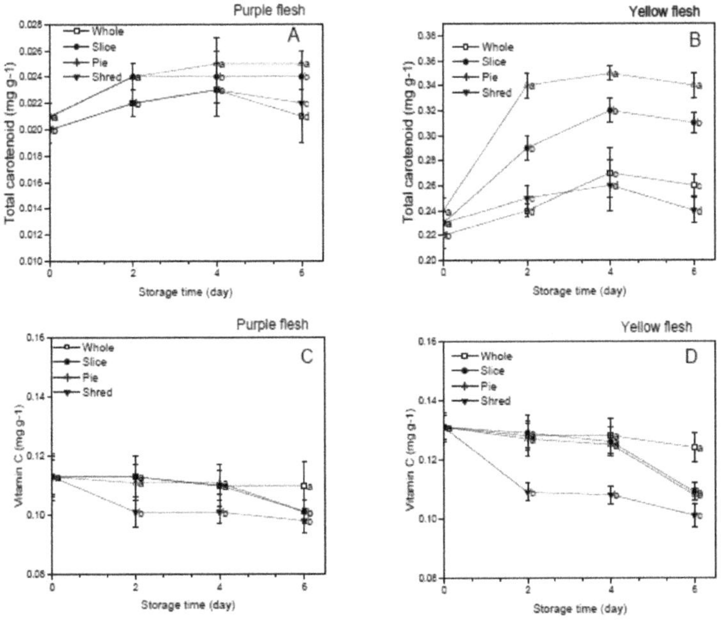

Figure 3. Effect of cutting styles on the total carotenoid (**A,B**), vitamin C (**C,D**) content of the purple and yellow flesh the sweet potato cultivars, respectively.

3.3. Effect of Cutting Styles on the Total Phenolics and Flavonoid Content

The effect of cutting styles on the total phenolics and flavonoid contents of the purple and yellow flesh sweet potato cultivars stored at 4 °C is presented in Figure 2A–D. The results indicated that the cutting significantly ($p < 0.05$) increased the total phenolics and flavonoid accumulation in both cultivars compared to the whole sweet potatoes. Wounding stress has been shown to increase phenolic accumulation by activating the phenylpropanoid pathway in fresh-cut fruit and vegetables [4,7]. Similar results were also found in our research. The contents of the phenolics and flavonoid contents increased with storage time firstly and then decreased slightly after four days of the storage. The decrease of total phenolic could be explained by the increase in utilization rate and decrease in synthesis rate. The accumulation of phenolics and flavonoids in sliced and pie-cut sweet potatoes was significantly ($p < 0.05$) higher than that in the shredded sample in both cultivars, which could be related to the oxidation and utilization of these bioactive compounds in high wounding intensity samples. In sweet potatoes, the total phenolic content in shreds cutting samples showed decrease trends while that in slices and pie-cuts increased [24]. Phenolic compounds have been severally reported as the main contributor to the antioxidant capacity of plants [9] while flavonoids have been shown to have radical scavenging or chelating activities [25]. The phenolics and antioxidant capacity in plants have been reported to be influenced by cultivars, maturity, and other environmental factors such as sunlight exposure [26]. In our results, the purple flesh sweet potatoes presented higher total phenolics and flavonoid contents than the yellow flesh sweet potato cultivars.

3.4. Effect of Cutting Styles on the Total Carotenoid and Vitamin C Content

The effect of the cutting styles on the carotenoid and vitamin C contents of the purple and yellow flesh sweet potato cultivars is shown in Figure 3A–D. Sweet potato cultivars are rich in carotenoid, particularly beta-carotene responsible for conferring pro-vitamin A activity that contributes to the prevention of vitamin A deficiencies and night blindness [10]. Furthermore, carotenoids are bioactive secondary metabolites in the plants that have been linked to health protection in in-vitro, in-vivo, and clinical research [27]. The study demonstrated that the carotenoid content in the yellow flesh sweet potatoes was higher than that in the purple flesh sweet potatoes, and cutting significantly ($p < 0.05$) increased the total carotenoid content compared to the whole sweet potatoes in both the purple and yellow flesh sweet potato cultivars. The total carotenoid content increased steadily among the cuts but declined slightly after four days of storage in both cultivars.

The result further revealed that the pie-cut had the highest total carotenoid content, while the shredded had the lowest carotenoid content in both cultivars throughout the storage period. This finding further indicated that the excessive wounding intensity in the shredded sample adversely affected the rate of wound-induced carotenoid biosynthesis. These results implied that the pie cutting of the sweet potato cultivars made it a better source of carotenoid and improved its bioactive properties compared to the whole sweet potatoes. This finding corroborated the previous report of Grace et al. [10] in which carotenoid biosynthesis of sweet potato cultivars was reported to increase with the storage time. However, unlike the carotenoid content, the vitamin C content of the purple and yellow flesh sweet potato decreased significantly ($p < 0.05$) in the shredded sample while the sliced and the pie-cut was not significantly ($p < 0.05$) affected until after four days of storage. The result of this study further supports the previous finding of Huang et al. [16] in who reported that the vitamin C content of stored sweet potato cultivars decreased with increased storage time. At the end of the storage time, the vitamin C content in the fresh-cut potatoes was lower than the whole, and the vitamin C content in the shredded sample was the least in both cultivars, which could be related to the damage of cell and the oxidation of vitamin C caused by high wounding intensity. Vitamin C is an essential part of a daily diet, and a decrease in vitamin C content, as seen in the shredded sample, implies lower nutritional quality, which would affect its consumer acceptability.

3.5. Effect of Cutting Styles on the DPPH Free Radical Scavenging Activity and Ferric Reducing Antioxidant Power

High antioxidant activity in food materials had been extensively studied and shown to correlates an increase in cardio-protective, hepato-protective, antidiabetic, as well as other physiological functions in both in-vitro and clinical studies [7,28]. DPPH free radical scavenging activity and ferric reducing antioxidant power (FRAP) were used to assess the effect of the cutting styles on the antioxidant ability of the sweet potato cultivars. Fresh-cut processing significantly ($p < 0.05$) increased the DPPH free radical scavenging activity and FRAP in both the purple and yellow flesh sweet potatoes compared to the whole sweet potato cultivar during the storage period. The pie produce had the highest radical scavenging activity and ferric reducing antioxidant power among the fresh cut in both cultivars. However, the purple flesh sweet potato showed higher antioxidant activity compared to the yellow flesh sweet potato Figure 4A,B.

This could be due to higher phenolic and flavonoid content in the purple flesh sweet potato compared to the yellow flesh sweet potato. Antioxidant and free radical scavenging ability in sweet potato are largely attributed to phenolic content [28]. Previous research findings have demonstrated that flavonoids can significantly contribute to improving the radical scavenging activity of sweet potato cultivars [26].

There was strong positive correlation between the antioxidant activity and the phenolics, flavonoids, carotenoids, and vitamin C content of both the purple and yellow sweet potato cultivars (Tables 2 and 3). The total phenolics and flavonoids content showed significant ($p \leq 0.05$) positive correlation with the ferric reducing antioxidant power activity while total carotenoid significantly ($p \leq 0.01$) correlated with the FRAP activity. This further supports other research reporting that increases in phenolic and flavonoid content enhance the antioxidant activity of food material [7]

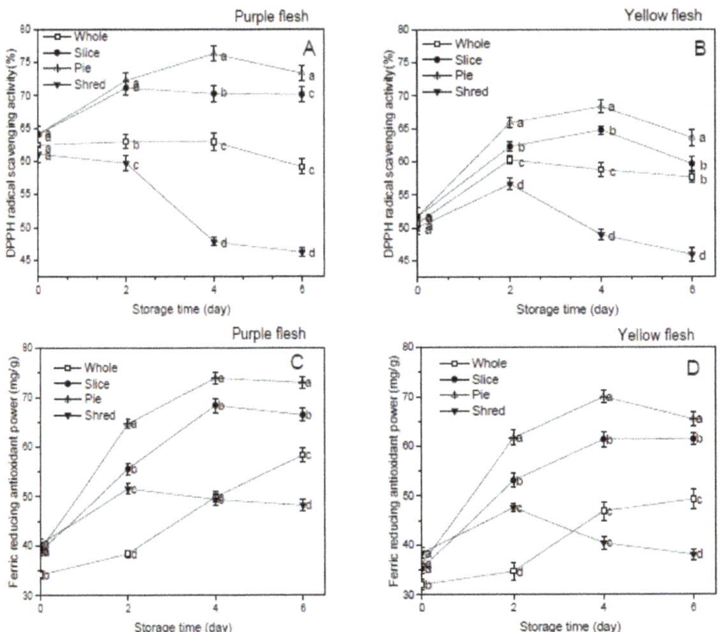

Figure 4. Effect of cutting styles on the DPPH radical scavenging activity (**A,B**) and ferric reducing antioxidant power (**C,D**) of the purple and yellow flesh sweet potato cultivars, respectively.

Table 2. Pearson correlation between the antioxidant activity (DPPH) and the chemical composition of the purple and yellow flesh sweet potato cultivars.

	Purple Flesh Sweet Potato						Yellow Flesh Sweet Potato				
	DPPH	Total Phenolics	Total Flavonoids	Total Carotenoids	Vitamin C		DPPH	Total Phenolics	Total Flavonoids	Total Carotenoids	Vitamin C
DPPH	1					DPPH	1				
Total phenolics	0.89898	1				Total phenolics	0.73097	1			
Total flavonoids	0.91732	0.96331 *	1			Total flavonoids	0.96981	0.86526	1		
Total carotenoids	0.86532	0.99568 *	0.96644 *	1		Total carotenoids	0.86865	0.952 *	0.96324 *	1	
Vitamin C	0.65444	0.26038	0.33792	0.18736	1	Vitamin C	0.71281	0.0704	0.52025	0.27173	1

* Means significant correlation at $p \leq 0.05$.

Table 3. Pearson correlation between the antioxidant activity (FRAP) and the chemical composition of the purple and yellow flesh sweet potato cultivars.

	Purple Flesh Sweet Potato						Yellow Flesh Sweet Potato				
	FRAP	Total Phenolics	Total Flavonoids	Total Carotenoids	Vitamin C		FRAP	Total Phenolics	Total Flavonoids	Total Carotenoids	Vitamin C
FRAP	1					FRAP	1				
Total phenolics	0.9857 *	1				Total phenolics	0.97686 *	1			
Total flavonoids	0.97557 *	0.96331 *	1			Total flavonoids	0.94999 *	0.86526	1		
Total carotenoids	0.99631 **	0.99568 **	0.96644 *	1		Total carotenoids	0.99393 **	0.952 *	0.96324 *	1	
Vitamin C	0.16286	0.26038	0.33792	0.18736	1	Vitamin C	0.24224	0.0704	0.5202 *5	0.27173	1

* Means significant correlation at $p \leq 0.05$ while ** means significant correlation at $p \leq 0.01$.

3.6. Polyphenol Oxidase (PPO) Activity

The PPO activity of the fresh-cut sweet potato cultivars is presented in Figure 5. PPO activity in potatoes increased with increased storage time, and similar results were also found in apples [8]. The PPO enzyme activity of the whole sweet potatoes was significantly lower ($p < 0.05$) compared to all the fresh-cut samples in both purple and yellow flesh sweet potato cultivars. The shredded sample showed the highest PPO activity compared to the sliced and pie-cuts in both cultivars. This implies that increasing wounding intensity in the potatoes sample enhanced the activity of PPO in both cultivars. PPO is the enzyme that is mainly involved in the browning process in injured plant tissues [8].

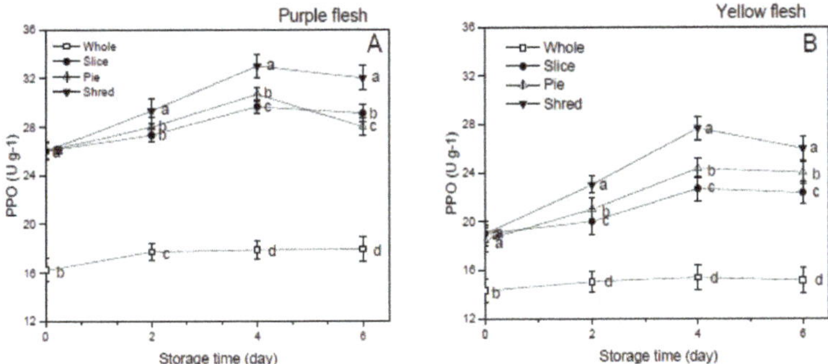

Figure 5. Effect of cutting styles on the PPO activity (**A,B**) of the purple and yellow flesh sweet potato cultivars, respectively.

PPO catalyzes the hydroxylation of mono and di-phenols to o-diphenols, which further oxidizes to o-quinones. These o-quinones condense and polymerize with certain amino acids and proteins to produce the undesirable brown/dark melanin pigments seen in fresh-cut fruits and vegetables [29]. Torres-Contreras et al. [24] reported that decreased phenolic content in shredded-potatoes during storage was linked to increased PPO activity. In our results, shredding the sweet potato cultivars induced the highest PPO and oxidative browning activity and which resulted in the increased brown surface color of the shredded sample compared to the sliced, pie-cuts, and the whole sweet potatoes (Figure 1 and Table 1). The subsequent oxidative browning induced by increased PPO activities in shred potatoes was undesirable in foods and food systems because of its changes in food appearance, development of off-flavors, and losses of nutrition quality [29,30]. This report is similar to the findings of Jang et al. [8] in which PPO activity increased with increased storage time and decreased consumer acceptability of fresh-cut apple.

3.7. Total Aerobic Bacterial Count (TABC)

The total aerobic count was determined to evaluate the effect of different cutting styles on the microbiological quality of the sweet potato cultivars (Figure 6). Fresh-cut produce is prone to decay by spoilage and pathogenic microbes that of public health significance [3]. In this study, TABC in potatoes increased with storage time and cutting induced the growth of microorganisms. Shredded sweet potatoes presented the most TABC in the four groups of samples in both cultivars, and it was significantly ($p < 0.05$) higher in the fresh-cut sweet potato cultivars than the whole sweet potatoes. The high-water activity and approximately neutral tissue pH in sweet potato cultivars promoted rapid microbial growth, which has been reported to contaminate fresh-cut produce and result in the faster deterioration of fresh-cut produce compared to whole fruits or vegetables [31]. Fresh-cut processing destroys plant tissue and exposes the nutrients rich cytoplasm to microbial contamination and spoilage [22].

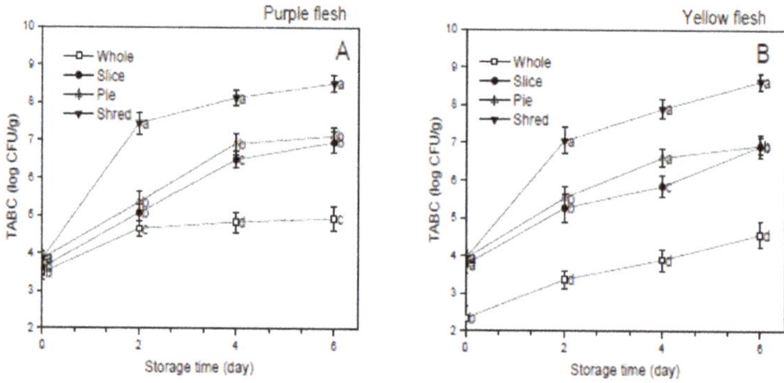

Figure 6. Effect of cutting styles on the total aerobic count (**A,B**) of the purple and yellow flesh the sweet potato cultivars, respectively.

Hence the finding of this study is similar to the results of other studies on the microbiological quality of fresh-cut produce [3,31]. However, the pie and sliced samples presented significantly ($p < 0.05$) less microbial numbers among the fresh-cut produce and suggested they would be better alternatives in fresh-cut processing of sweet potato cultivars to shredding. Food safety constitutes a growing concern for producers, public, and relevant regulatory agencies in the fresh-cut processing industry; hence, selecting cutting styles that ensure minimal contamination of the fresh-cut produce remains a critical focus for all relevant stakeholders in the fresh-cut processing industry [31]. The present investigation suggested that pie cutting was a useful tool in improving the bioactive contents as well as the antioxidant activity of potatoes.

4. Conclusions

In this study, the effect of cutting styles on the quality characteristics and antioxidant activity of sweet potato cultivars was evaluated. The results indicated that cutting could enhance the antioxidant activity of sweet potatoes, but shred cutting had the adverse effect on the quality of the sweet potatoes due to its excessive wounding stress. The pie samples exhibited the highest total phenolic, flavonoid, and carotenoid accumulation and thus presented the highest antioxidant activity in both cultivars. The PPO activity was excessively high in the shredded sample and resulted in the decrease of phenolic antioxidants and the change of color. Therefore, the finding of the current study demonstrated that it is possible to provide the consumer with fresh-cut sweet potatoes rich in bioactive compounds by simply pie cutting, which shows potential in the processing of sweet potatoes.

Author Contributions: Conceptualization, A.K.D. and P.J.; Data curation, A.K.D. and S.U.F.B.; Funding acquisition, P.J.; Investigation, A.K.D., L.W., S.U.F.B. and M.P.M.; Methodology, A.K.D. and L.W.; Supervision, P.J.; Validation, P.J. and Y.Z.; Visualization, M.P.M.; Writing—original draft, A.K.D.

Funding: This study was supported by the Project Funded by the Priority Academic Program Development of Jiangsu Higher Education Institutions and Program for Student Innovation Through Research and Training (SRT) (No. 1918C12).

Conflicts of Interest: The authors declare no conflict of interest.

References

1. Guo, K.; Liu, T.; Xu, A.; Zhang, L.; Bian, X.; Wei, C. Structural and functional properties of starches from root tubers of white, yellow, and purple sweet potatoes. *Food Hydrocoll.* **2019**, *89*, 829–836. [CrossRef]
2. Hu, Y.; Deng, L.; Chen, J.; Zhou, S.; Liu, S.; Fu, Y.; Yang, C.; Liao, Z.; Chen, M. An analytical pipeline to compare and characterize the anthocyanin antioxidant activities of purple sweet potato cultivars. *Food Chem.* **2016**, *194*, 46–54. [CrossRef] [PubMed]

3. Qadri, O.S.; Yousuf, B.; Srivastava, A.K. Fresh-cut fruits and vegetables: Critical factors influencing microbiology and novel approaches to prevent microbial risks-A review. *Cogent Food Agric.* **2015**, *1*, 1121–1606. [CrossRef]

4. Li, X.; Long, Q.; Gao, F.; Han, C.; Jin, P.; Zheng, Y. Effect of cutting styles on quality and antioxidant activity in fresh-cut pitaya fruit. *Postharvest Biol. Technol.* **2017**, *124*, 1–7. [CrossRef]

5. Surjadinata, B.B.; Cisneros-Zevallos, L. Biosynthesis of phenolic antioxidants in carrot tissue increases with wounding intensity. *Food Chem.* **2012**, *134*, 615–624. [CrossRef]

6. Jacobo-Velázquez, D.A.; González-Agüero, M.; Cisneros-Zevallos, L. Cross-talk between signaling pathways: The link between plant secondary metabolite production and wounding stress response. *Sci. Rep.* **2015**, *5*, 8608. [CrossRef]

7. Jacobo-Velázquez, D.A.; Martínez-Hernández, G.B.; Rodríguez, S.D.C.; Cao, C.M.; Cisneros-Zevallos, L. Plants as biofactories: Physiological role of reactive oxygen species on the accumulation of phenolic antioxidants in carrot tissue under wounding and hyperoxia stress. *J. Agric. Food Chem.* **2011**, *59*, 6583–6593. [CrossRef]

8. Jang, J.H.; Moon, K.D. Inhibition of polyphenol oxidase and peroxidase activities on fresh-cut apple by simultaneous treatment of ultrasound and ascorbic acid. *Food Chem.* **2011**, *124*, 444–449. [CrossRef]

9. Han, C.; Li, J.; Jin, P.; Li, X.A.; Wang, L.; Zheng, Y.H. The effect of temperature on phenolic content in wounded carrots. *Food Chem.* **2017**, *215*, 116–123. [CrossRef]

10. Grace, M.H.; Yousef, G.G.; Gustafson, S.J.; Truong, V.D.; Yencho, G.C.; Lila, M.A. Phytochemical changes in phenolics, anthocyanins, ascorbic acid, and carotenoids associated with sweet potato storage and impacts on bioactive properties. *Food Chem.* **2014**, *145*, 717–724. [CrossRef]

11. Kim, H.W.; Kim, J.B.; Cho, S.M.; Chung, M.N.; Lee, Y.M.; Chu, S.M.; Che, J.H.; Kim, S.N.; Kim, S.Y.; Cho, Y.S.; et al. Anthocyanin changes in the Korean purple-fleshed sweet potato, Shinzami, as affected by steaming and baking. *Food Chem.* **2012**, *130*, 966–972. [CrossRef]

12. Tang, Y.; Cai, W.; Xu, B. Profiles of phenolics, carotenoids and antioxidative capacities of thermal processed white, yellow, orange and purple sweet potatoes grown in Guilin, China. *Food Sci. Hum. Wellness* **2015**, *4*, 123–132. [CrossRef]

13. Wall, M.M. Storage quality and composition of sweet potato roots after quarantine treatment using low doses of x-ray irradiation. *Hortic. Sci.* **2005**, *40*, 424–427.

14. Waterhouse, A.L. Determination of Total Phenolics. *Curr. Protoc. Food Anal. Chem.* **2011**, *6*, I1.1.1–I1.1.8.

15. Sun, B.; Ricardo-da-Silva, J.M.; Spranger, I. Critical factors of vanillin assay for catechins and proanthocyanins. *J. Agric. Food Chem.* **1998**, *46*, 4267–4274. [CrossRef]

16. Huang, Y.C.; Chang, Y.H.; Shao, Y.Y. Effects of genotype and treatment on the antioxidant activity of sweet potato in Taiwan. *Food Chem.* **2006**, *98*, 529–538. [CrossRef]

17. Arakawa, N.; Tsutsumi, K.; Sanceda, N.G.; Kurata, T.; Inagaki, C. A rapid and sensitive method for the determination of ascorbic acid using 4,7-diphenyl1,10-phenanthroline. *Agric. Biol. Chem.* **1981**, *45*, 1289–1290.

18. Bae, S.H.; Suh, H.J. Antioxidant activities of five different mulberry cultivars in Korea. *LWT-Food Sci. Technol.* **2007**, *40*, 955–962. [CrossRef]

19. Chen, C.C.; Lin, C.; Chen, M.H.; Chiang, P.Y. Stability and Quality of Anthocyanin in Purple Sweet Potato Extracts. *Foods* **2019**, *8*, 393. [CrossRef]

20. Manohan, D.; Wai, W.C. Characterization of polyphenol oxidase in sweet potato (*Ipomoea batatas* L.). *J. Adv. Sci. Arts* **2012**, *3*, 14–31.

21. Soison, B.; Jangchud, K.; Jangchud, A.; Harnsilawat, T.; Piyachomkwan, K. Characterization of starch in relation to flesh colors of sweet potato varieties. *Int. Food Res. J.* **2015**, *22*, 2302–2308.

22. Barry-Ryan, C.; Pacussi, J.M.; O'Beirne, D. Quality of shredded carrots as affected by packaging film and storage temperature. *J. Food Sci.* **2000**, *65*, 726–730. [CrossRef]

23. Huang, C.L.; Liao, W.C.; Chan, C.F.; Lai, Y.C. Storage performance of Taiwanese sweet potato cultivars. *J. Food Sci. Technol.* **2014**, *51*, 4019–4025. [CrossRef]

24. Torres-Contreras, A.M.; Nair, V.; Cisneros-Zevallos, L.; Jacobo-Velázquez, D.A. Plants as biofactories: Stress-induced production of chlorogenic acid isomers in potato tubers as affected by wounding intensity and storage time. *Ind. Crop. Prod.* **2014**, *62*, 61–66. [CrossRef]

25. Hue, S.M.; Boyce, A.N.; Somasundram, C. Antioxidant activity, phenolic and flavonoid contents in the leaves of different varieties of sweet potato ('ipomoea batatas'). *Aust. J. Crop Sci.* **2012**, *6*, 375–380.

26. Teow, C.C.; Truong, V.D.; McFeeters, R.F.; Thompson, R.L.; Pecota, K.V.; Yencho, G.C. Antioxidant activities, phenolic, and β-carotene contents of sweet potato genotypes with varying flesh colours. *Food Chem.* **2007**, *103*, 829–838. [CrossRef]

27. Tomlins, K.; Owori, C.; Bechoff, A.; Menya, G.; Westby, A. Relationship among the carotenoid content, dry matter content and sensory attributes of sweet potato. *Food Chem.* **2012**, *131*, 14–21. [CrossRef]

28. Rabah, I.O.; Hou, D.X.; Komine, S.I.; Fuji, M. Potential chemopreventive properties of extract from baked sweet potato (Ipomoea batatas Lam. Cv. Koganesengan). *J. Agric. Food Chem.* **2004**, *52*, 7152–7157. [CrossRef]

29. Arogundade, L.A.; Mu, T.H. Influence of oxidative browning inhibitors and isolation techniques on sweet potato protein recovery and composition. *Food Chem.* **2012**, *134*, 1374–1384. [CrossRef]

30. Severini, C.; Baiano, A.; De Pilli, T.; Romaniello, R.; Derossi, A. Prevention of enzymatic browning in sliced potatoes by blanching in boiling saline solutions. *LWT–Food Sci. Technol.* **2003**, *36*, 657–665. [CrossRef]

31. Harris, L.J.; Farber, J.N.; Beuchat, L.R.; Parish, M.E.; Suslow, T.V.; Garrett, E.H.; Busta, F.F. Outbreaks associated with fresh produce: Incidence, growth, and survival of pathogens in fresh and fresh-cut produce. *Compr. Rev. Food Sci. Food Saf.* **2003**, *2*, 78–141. [CrossRef]

 © 2019 by the authors. Licensee MDPI, Basel, Switzerland. This article is an open access article distributed under the terms and conditions of the Creative Commons Attribution (CC BY) license (http://creativecommons.org/licenses/by/4.0/).

Article

Combination of Low Fluctuation of Temperature with TiO₂ Photocatalytic/Ozone for the Quality Maintenance of Postharvest Peach

Xiaoyu Jia [1], Jiangkuo Li [2], Meijun Du [1], Zhiyong Zhao [3], Jianxin Song [1], Weiqiao Yang [4,5], Yanli Zheng [1], Lan Chen [1] and Xihong Li [1,*] 

[1] State Key Laboratory of Food Nutrition and Safety, College of Food Science and Engineering, Tianjin University of Science and Technology, Tianjin 300457, China; jiaxiaoyu331@outlook.come (X.J.); dumeijun2018@163.com (M.X.); sjxtust@126.com (J.S.); zhengyanli3344@163.com (Y.Z.); chenlan890804@163.com (L.C.)

[2] Tianjin Key Laboratory of Postharvest Physiology and Storage of Agricultural Products, National Engineering and Technology Research Center for Preservation of Agricultural Products, Tianjin 300384, China; lijkuo@sina.com

[3] Instiute of Agro-Products Processing Science and Technology, Xinjiang Academy of Agricultural and Reclamation Science, Shihezi 832000, China; yll520520520520@sina.com

[4] Academy of National Food and Strategic Reserves Administration, Beijing 100037, China; yangweiqiaospring@126.com

[5] Tianjin Gasin-DH Preservation Technologies Co., Ltd., Tianjin 300403, China

* Correspondence: lixihong@tust.edu.cn; Tel.: +86-022-6091-2406

Received: 20 January 2020; Accepted: 20 February 2020; Published: 21 February 2020

Abstract: Chilling injury, tissue browning, and fungal infection are the major problems of peach fruit during post-harvest storage. In this study, a precise temperature control cold storage with low-temperature fluctuation (LFT) and internal circulation flow system is designed. An ozone (O_3) generator and a (titanium dioxide) TiO_2 photocatalytic reactor were applied to cold storage to investigate the variation of LFT combined with ozone fumigation and a TiO_2 photocatalytic reactor in the efficiency of delaying ripening and maintaining peach fruit quality. Results showed that the temperature fluctuation with the improved control system was only ±0.1 to ±0.2 °C compared with that of ±0.5 to ±1.0 °C in conventional cold storage. LFT significantly reduced the chilling injury of peach fruit during storage. Although LFT combined with fumigation of 200 mg m^{-3} ozone periodical treatment slightly damaged the peach fruit after 40 d of storage, its combination with the TiO_2 photocatalytic system significantly improved the postharvest storage quality of the fruit. This treatment maintained higher titratable acidity (TA), total soluble solids (TSS), better firmness, color, microstructure, and lower decay rate, polyphenol oxidase (PPO) activities, total phenol accumulation, respiratory intensity, ethylene production, and malondialdehyde (MDA) content during 60 d of storage. All the results show that LFT combined with the TiO_2 photocatalytic system might be a promising technology for quality preservation in peach fruit storage.

Keywords: peach; chilling injury; internal circulation system; low fluctuation of temperature; TiO_2 photocatalytic; storage quality

---

1. Introduction

Peach [*Prunus persica* (L.) Batsch] belongs to the Rosaceae family, with a long history of cultivation in China. The output quantity of peach is more than 8 million tons every year [1]. It is worth noting that peach presents strong respiration intensity and rapid softening after harvest, which leads to short shelf-life [2]. Up to now, various techniques such as normal atmosphere cold storage, hypobaric storage, gamma-irradiation, heat treatments, and controlled atmosphere cold storage have been applied in peach preservation [3–5]. Among these technologies, cold storage is the most commonly used, especially for large amounts of peaches. However, during low-temperature storage, chilling injuries, quick softening, browning, woolly texture, and high perishability are still the main problems [6].

Accurate control of temperature and humidity in a cold storage room is generally not simple [7]. Non-uniform airflow from air coolers, evaporator defrosting processing, and frequent door opening and closing during storage result in the fresh products being exposed to undesired high-temperature fluctuations [8]. At present, temperature fluctuations in commercial cold storage rooms are approximately ±0.5 to ±1.0 °C [9]. Peach, like cucumber and mango, is sensitive to temperature fluctuation and prone to chilling injury during storage [10]. Chilling injury is mainly caused by the conversion of membrane lipids from liquid crystal phase to rigid solid gel phase, which destroys the integrity of the cell membrane [11]. Temperature also affects respiration and postharvest rot rates of the product [12]. The application of a jacketed storage system provided a feasible scheme to overcome these disadvantages, in which refrigerated air is circulated through an air space surrounding the storage room rather than within the space itself [13]. Although the construction cost of jacketed storage is 15% higher than that of conventional cold storage, the use of jacketed systems can significantly improve product quality and storage life, which depends on better temperature control and higher relative humidity [14–17]. In addition, the jacketed system can significantly reduce energy consumption because of sensible cooling during the dormant storage period and less defrosting times [13].

Titanium dioxide (TiO_2) and ozone have been evaluated as promising sanitizers for fresh fruit and vegetables [18,19]. As a wide band gap (3.2 eV) semiconductor under ultroviolet (UV) (320–400 nm) illumination, TiO_2 generates energy-rich electron-hole pairs that can be transferred to the surface of TiO_2 and promotes reactivity with the surface-absorbed molecules leading to the production of active radicals [20]. TiO_2 efficiently promotes the photocatalytic oxidation of organic compounds and the oxidation of microorganism cell membranes [21]. In practical application, TiO_2 is mainly used in food packaging and photocatalytic reactors [19,22]. TiO_2 photocatalytic reactors had been applied to tomato fruit during storage to delay ripening time [23]. Because of its ability to reduce microorganisms and for the oxidation of ethylene, gaseous ozone had been successfully applied for the storage of various fruits such as apples, papayas, orange potatoes, pears, and strawberries [24,25]. However, there was no research into the effect of LFT combined with ozone and a TiO_2 photocatalytic system on postharvest quality of peaches.

Although both ozone and TiO_2 photocatalysis treatments have been studied in a variety of fruit and vegetables, little has been done in their combination with low-temperature fluctuation for peach storage. Here, the improved temperature control and the combination of this technique with TiO_2 photocatalysis or ozone intermittent treatment were investigated in the cold storage of peaches, and important quality parameters such as TSS, (polyphenol oxidase) PPO activity, total phenol content, (malondialdehyde) MDA, and fruit microstructure were analyzed in cold-stored peaches. The main purpose of this study is to evaluate the optimal temperature and post-harvest quality control techniques for cold storage of peaches.

2. Materials and Methods

2.1. Plant Material

Peaches [Prunus persica (L.) Batsch cv. Jin Qiu hong] were obtained from an orchard in Beijing, China. Only fruit free from damage, diseases or infestations, and of approximately a uniform size (250 to 300 g) and maturity were selected and pre-cooled to 5 °C [26].

2.2. Specification of Cold Storage

The structure of conventional cold storage is shown in Figure 1A. The cold storage with improved precise temperature control is composed of both a separate external jacket and an internal storage room, refrigeration equipment, and an internal circulation flow system. The jacket-room (6700 × 6200 × 4600 mm) was a conventional cold store (Figure 1B), while the internal storage room (6000 × 5000 × 3500 mm) was made up of 1.5-mm thick aluminum plates. The fruit were placed in the internal room.

2.3. Gaseous Ozone and TiO$_2$ Treatment

The ozone and TiO$_2$ photocatalytic reactor were separately installed in the internal circulation system (Figure 1C,D). Ozone (90%) was produced by an FL-803Y ozone generator (Shenzhen Feili Electrical Technology Co. Ltd., Shenzhen, China). The JSDHMK-1227 TiO$_2$ photocatalytic reactor was provided by a local company (Tianjin Gasin-DH Preservation Technologies Co. Ltd., Tianjin, China). The particle specific surface area and equivalent particle size of TiO$_2$ were 11.7 m^2 g^{-1} and 93.7 nm, respectively, the wavelength of ultraviolet light was 365 nm. The TiO$_2$ coated surface area was 592.8 cm^2.

2.4. Storage Condition

Peach samples were randomly divided into 40 plastic baskets (10 kg per basket). Peaches of each treatment (10 baskets) were transferred to the corresponding cold rooms, which were set as 0 °C. All cold storage rooms were built by Tianjin Gasin-DH Preservation Technologies Limited Co. Ltd (Tianjin, China). The treatments were as follows: CK (control), stored in a conventional cold storage room; LFT, stored in a precise temperature control cold storage; LFT + O$_3$, stored in a precise temperature control cold storage and fumigated with ozone concentration of 200 mg m^{-3} (selected from previous experiments) for 30 min every week; LFT + TiO$_2$, stored in a precise temperature control cold storage and treated with a TiO$_2$ photocatalytic system for 30 min every week. The relative humidity for all treatments was 90% ± 5%. For each treatment, the experimental measurements were taken every ten days.

2.5. Analysis of Cold Storage Temperature Fluctuation

A BTC-A16 temperature and humidity recording instrument (Tianjin Boyuanda Technology Co. Ltd., Tianjin, China) was used to measure the temperature fluctuation in all storage rooms. The uniformity of temperature distribution in the conventional and the improved cold storage was evaluated at 35 temperature measurement points at heights of 0.5 and 3.5 m.

2.6. Analysis of Physicochemical Properties

2.6.1. Color and Firmness Analysis

Fruit surface color was determined on three different locations of each individual fruit using the HP-200 colorimeter (Shanghai Chinaspec Optoelectronics Technology Co. Ltd., Guangzhou. China) [27]. The L* (lightness), a* (reddish-greenish) and b* (yellowish-bluish) indexes of the International Commission on Illumination Lab color space (CIELAB) colorimetric system were used to evaluate the color change of the peach samples [28].

The firmness of flesh fruit was measured using a GY-4 digital penetrometer (Zhejiang Top Instrument Co. Ltd., Hangzhou, China) with the needle-like probe of 10 mm diameter [29]. The results were expressed as Newton (N). Ten replicates were measured for this analysis, in which two fruit were used for each replicate.

2.6.2. Titratable Acidity (TA) and Total Soluble Solids (TSS) Analysis

TA was measured by acid-base titration and expressed as a malic acid content percentage [30]. The total soluble solid (TSS) content of a peach was measured by a PAL-1 refractometer (Atago Co. Ltd., Tokyo, Japan) [31]. Ten replicates were performed for this analysis, in which two fruit were used for each replicate.

2.6.3. Respiratory Rate and Ethylene Production

A GXH-305 infrared gas analyzer (Junfang Science & Technology Institute of Physical and Chemical Research, Beijing, China) was used to measure the respiratory rate, according to the method of Yang et al. [32]. Peach samples were placed in gas-tight jars for 1 h at 0 °C. Ethylene production was measured according to the method of Huan et al. [33]. The head space gas (1 mL) gas was injected into an Agilent GC 7890 gas chromatograph (Agilent Technologies, Santa Clara, CA, USA) equipped with a flame ionization detector (FID). Fresh mass-based rates of respiration were measured as CO_2 release ($mg\ kg^{-1}\ h^{-1}$) and those of ethylene production were given as $\mu L\ kg^{-1}\ h^{-1}$.

2.6.4. Decay Rate Analysis

The decay rate was measured by a previous method with visual evaluation [34]. The growth of mold on peach was regarded as decay, and the analysis carried out 10 repetitions based on 25 peaches.

2.6.5. Polyphenol Oxidase (PPO) Activity and Total Phenolics Content Analysis

PPO activity was measured according to the method of Wang et al. [28]. Briefly, 3.0 g tissue was homogenized in 10 mL of 0.05 mol L^{-1} sodium phosphate buffer (pH 7.8) and 0.5 g of polyvinylpyrrolidone (PVPP). Then the homogenate was centrifuged in a 5804-r refrigerated centrifuge (Eppendorf, International Trade Co., Ltd., Shanghai, China) at 5000× g for 10 min. The supernatant (0.5 mL) was mixed with 1.5 mL phosphate buffer (0.05 mol L^{-1}, pH 7.8) and 1 mL catechol (0.1 mol L^{-1}). The change of optical density (OD) of the reaction mixture was measured every 30 s for 3 min at 420 nm by a 756-PC UV–vis spectrophotometer (T6 New Century, Beijing Purkinje General Instrument Co., Ltd., Beijing, China). The result of PPO activity was expressed as U g^{-1} fresh weight, where U = 0.01 Δ420 nm per min.

The Folin–Ciocalteu procedure was used to measure the total phenolic content according to the method of Piccolella et al. [35] with modifications. Briefly, 2.0 g of peach samples were crushed and added with 5.0 mL of 60% ethanol solution (*v/v*), then centrifuged at 3000× g for 10 min. Next, 0.25 mL of the supernatant was obtained and added with 5.0 mL of distilled water, 0.25 mL of Folin–Phenol reagent and 0.8 mL of 20% (*w/w*) Na_2CO_3, respectively. Then the mixture was placed in dark place for 30 min and measured at 760 nm absorbance. The results were expressed as g kg^{-1} fresh weight.

2.6.6. Malondialdehyde (MDA) Content Analysis

Malondialdehyde (MDA) content was measured by thiobarbituric acid-reactive substance [36]. Peach tissue (5.0 g) was homogenized with 10 mL of trichloroacetic acid (TCA, 100 g L^{-1}) and then centrifuged at 16,000× g for 10 min. The reaction mixture was a blend of 2.0 mL supernatant and 2.0 mL of thiobarbituric acid (0.5%, TBA), was heated in boiling water and cooled before centrifuged at 1000× g for 15 min. Finally, the absorbance of the mixture was measured at 532, 600, and 450 nm. MDA content was expressed on a fresh weight basis as mmol kg^{-1}.

2.6.7. Scanning Electron Microscope (SEM) Analysis

Microstructure of peach fruit was analysed with a SEM, as previously described [34]. The peach samples (approximately 3.0 × 3.0 × 1.0 mm) were cut from fruit with a blade. The samples were freeze-dried in a FD-1A-50 lyophilizer for 12 h (Shanghai billon instrument Co. Ltd., Shanghai, China) and then coated with 25 nm thick gold using a Balzers Union SCD 040 Sputter Coater (Balzer,

Wiesbaden, Germany) before SEM analysis. Representative areas were examined with a Hitachi S-2500 (Hitachi Ltd., Tokyo, Japan) scanning electron microscopy at an accelerating voltage of 20 kV.

2.7. Statistical Analysis

Experiments were performed in a completely randomized manner with 10 replicates. SPSS version 13.0 software (SPSS Inc., Chicago, IL, USA) was used for the one-way analysis of variance (ANOVA) at the level of $p < 0.05$. All data was repeated ten times and expressed as mean ± SD (standard deviation).

3. Results and Discussion

3.1. Temperature Fluctuation

Random temperature fluctuations can cause the centre temperature of chilled produce to decrease temporarily below the threshold level beyond which cold injury may develop [37]. Moreover, temperature fluctuations may exacerbate moisture condensation, which lead to microbial growth and fruit rot [38]. In order to analyze the temperature fluctuation of the cold storage system, the temperature distribution of 0.5 and 3.5 m height planes in the conventional cold storage and the improved cold storage was measured. At a temperature setting of 0 °C, the temperature fluctuation in the jacketed storage room was small at only ±0.1 to ±0.2 °C (Figure 2A,B), if compared to ±0.5 to ±1.0 °C in the conventional cold store (Figure 2C,D). The latter was temperature-controlled by the refrigeration equipment (Figure 1A). Direct cooling was used and the products were exposed to the cold blown air [13]. In the improved cold storage system, the storage room was cooled by air flowing through an enclosed space or jacket surrounding the walls, floor, and ceiling rather than by direct circulation of air through the room (Figure 1B,C). By adding an inner structure, it avoided direct contact between the products and the evaporator, and maintained a low temperature fluctuation in the internal room. A similar jacket system was reported by Raghavan et al. [18] and used to store fresh carrots, which inhibited the loss of moisture and maintained postharvest quality after long-term storage [17].

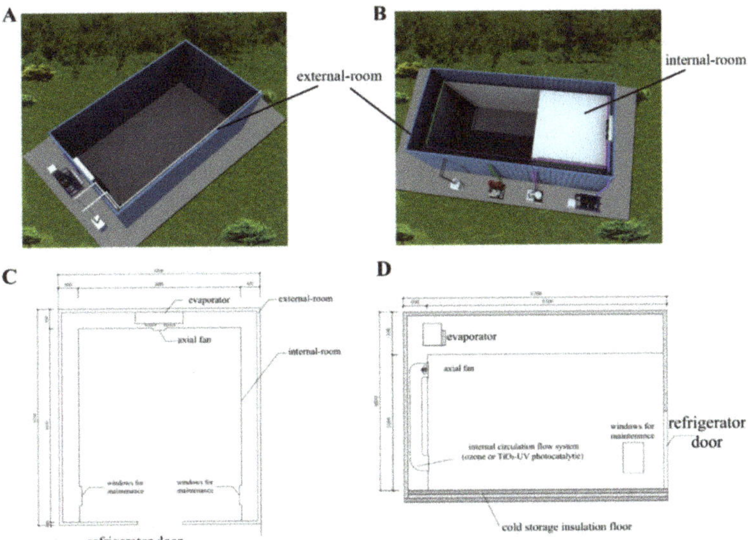

Figure 1. Comparison of conventional cold storage and LFT controlled cold storage. (**A**) Structure of conventional cold storage room. (**B**) Structure of LFT controlled cold storage room. (**C**) Floor plan of LFT cold storage room. (**D**) Horizontal profile of LFT controlled cold storage room.

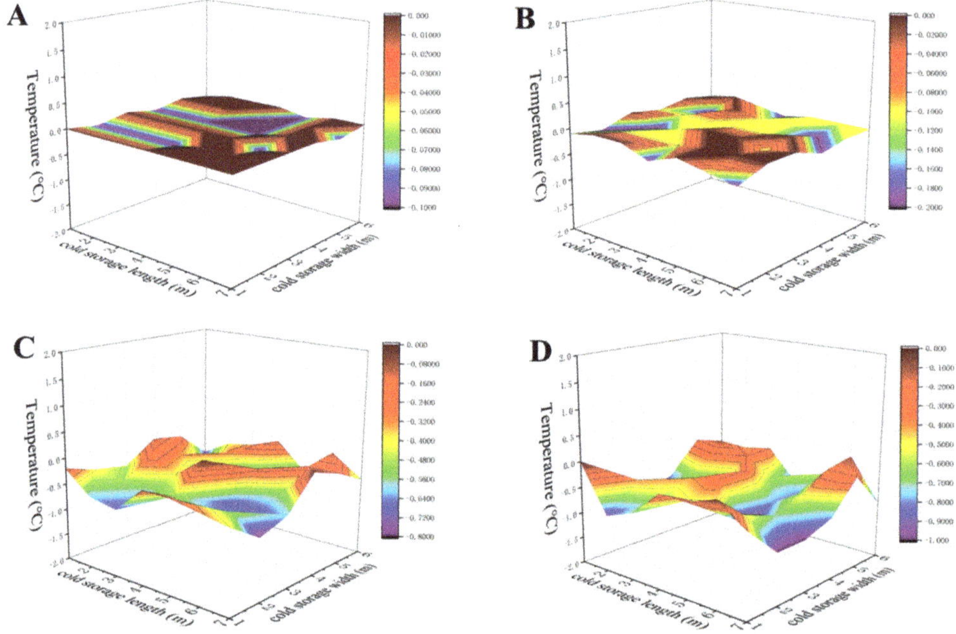

Figure 2. Temperature fluctuations at 0.5 (**A,C**) and 3.5 m (**B,D**) height in the improved (**A,C**) and the conventional (**B,D**) cold storage.

3.2. Physicochemical Properties

3.2.1. Color and Firmness

Figure 3 shows the appearance and section of peaches during storage for 30 and 60 d. For fruit from both cold store types, no significant differences in the color of the cut surfaces were found after 30 d of storage. In contrast, after 60 d, peaches of the CK group showed symptoms of severe chilling, i.e., browning, woolliness, and flesh translucency. The color of peach slices in LFT, LFT + O_3, and LFT + TiO_2 groups were all better than of those of the CK group, which indicates that LTF significantly reduced the chilling damage of peach fruit during storage. On the skin of LFT + O_3 peaches, symptoms of injury and pitted structures appeared after 60 d of storage. Similarly, high concentrations of gaseous ozone also caused damage to carrots [39]. Fruit of the LFT + TiO_2 group showed the best appearance and bright flesh color (Figure 3). In agreement with the appearance observations, the L* value gradually decreased, whereas the b* value gradually increased throughout the storage of 60 d (Table 1). L* is the lightness and corresponds to a darkbright scale (0, black; 100, white). After storage for 20 d, L* value of the LFT, LFT + O_3, and LFT + TiO_2 peaches were higher than those of the CK group, indicating that LFT combined with either ozone fumigation or TiO_2 photocatalysis could better maintain the L* value of peach fruit. Oddly, the L* value of peaches of the LFT + O_3 group was lower than those of the LFT+TiO_2 group after 40 d, which may be due to slight oxidation and damage of peach fruit caused by repeated ozone fumigation. A similar result was reported by Bridges et al. [40] where exposure to gaseous O_3 at 1.71 mg g^{-1} for 5.0 h resulted in noticeable bleaching of carrot and tomato tissue. The increase of b* value indicates the deepening of yellow. In the experiment, the values of b* in the CK group were significantly higher than those in other treatment groups after 30 d. The LFT + TiO_2 treatment showed the lowest b* value at 60 d, which expressed the lowest color change. Furthermore, the value of a* was not displayed due to no-significant differences between among different treatments.

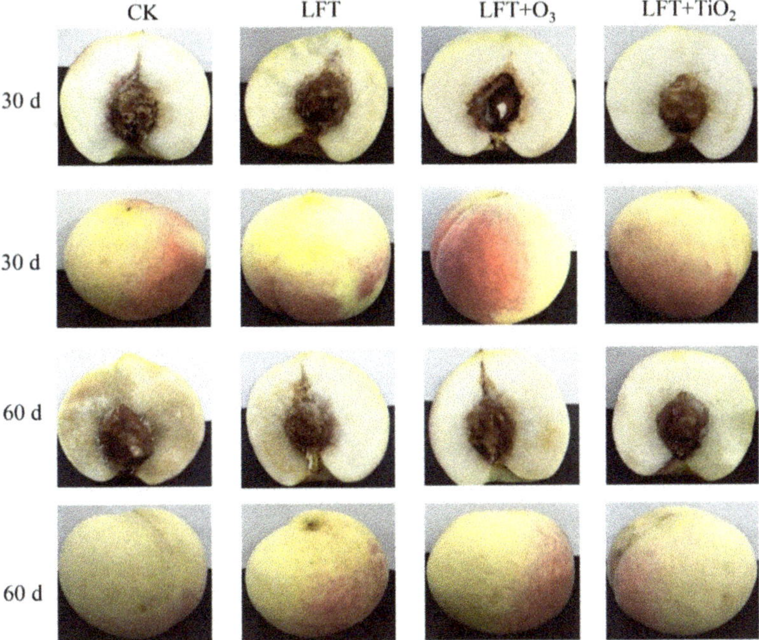

Figure 3. Effect of control (CK), LFT, LFT combined with ozone treatment (LFT + O$_3$) and LFT combined with TiO$_2$ photocatalysis (LFT + TiO$_2$) treatments on the appearance and longitudinal section photos of peach fruit after storage of 30 and 60 d.

The firmness of peach fruit in all groups continuously declining during storage; however, the firmness of fruits in LFT + O$_3$ and LFT + TiO$_2$ groups were significantly higher than that those of the CK and LFT groups after 40 d of storage ($p < 0.05$; Table 1). There was no significant difference between fruits in the LFT + O$_3$ group and LFT + TiO$_2$ group ($p > 0.05$). These results indicate that the combination of LFT and TiO$_2$ photocatalysis or ozone treatment significantly impacts fruit firmness, which might be related to degraded ethylene during storage.

3.2.2. Titratable Acidity (TA) and Total Soluble Solids (TSS)

TA values declined throughout the entire storage period for all treatments (Table 1). However, both LFT + O$_3$ and LFT+TiO$_2$ treatments delayed the loss of TA by 0.15% and 0.25%, respectively, during the 60 d of storage. Similar results were demonstrated by Ali et al. [41] and Li et al. [42] where papaya and strawberry fruit showed a decreasing tendency in TA value treated with ozone and TiO$_2$–LDPE packaging, respectively.

TSS content is an important indicator of fruit maturity and intrinsic quality. There were no significant ($p > 0.05$) differences among all groups initially, but TSS content in CK group declined rapidly after 30 d. The TSS content in the LFT + TiO$_2$ group increased from an initial value of 15.12% to 15.92% at 9 d of storage and then decreased to 13.20% at 60 d. However, due to the lack of a continuous supply of organic substances in the later period of storage, TSS can be consumed by respiration, resulting in the decline of TSS content [43]. At the end of storage (60 d), there was a significant difference among the TSS contents of the CK, LFT + TiO$_2$, and LFT + O$_3$ groups. These results were in accordance with the finding of Xing et al. [44] that the nano-TiO$_2$ coating treatment could maintain mango quality by delaying the decline in TSS content.

Table 1. Changes of physicochemical properties of different treatments after 60 d storage.

Physicochemical Properties	Storage Period (d)	Treatments			
		CK	LFT	LFT + O_3	LFT + TiO_2
Color (L* value)	0	78.75 ± 0.12 [a]	78.75 ± 0.09 [a]	78.82 ± 0.08 [a]	78.85 ± 0.05 [a]
	10	76.47 ± 0.10 [a]	76.49 ± 0.11 [a]	76.53 ± 0.07 [a]	76.81 ± 0.06 [a]
	20	73.44 ± 0.13 [c]	74.81 ± 0.10 [b]	76.35 ± 0.05 [a]	76.23 ± 0.06 [a]
	30	72.56 ± 0.11 [c]	74.20 ± 0.12 [b]	75.60 ± 0.04 [a]	74.86 ± 0.05 [ab]
	40	68.54 ± 0.09 [c]	72.35 ± 0.10 [b]	73.25 ± 0.13 [b]	74.56 ± 0.12 [a]
	50	65.52 ± 0.13 [d]	69.01 ± 0.09 [c]	71.52 ± 0.08 [b]	73.67 ± 0.12 [a]
	60	59.95 ± 0.10 [d]	63.79 ± 0.14 [c]	70.46 ± 0.10 [b]	72.43 ± 0.09 [a]
Color (b* value)	0	5.25 ± 0.02 [a]	5.16 ± 0.01 [a]	5.21 ± 0.01 [a]	5.23 ± 0.02 [a]
	10	5.66 ± 0.01 [a]	5.56 ± 0.02 [a]	5.53 ± 0.03 [a]	5.51 ± 0.04 [a]
	20	6.01 ± 0.04 [a]	5.92 ± 0.05 [a]	5.96 ± 0.03 [a]	5.94 ± 0.03 [a]
	30	6.32 ± 0.05 [a]	6.22 ± 0.02 [a]	6.28 ± 0.03 [a]	6.19 ± 0.02 [a]
	40	7.82 ± 0.01 [a]	6.86 ± 0.01 [b]	6.57 ± 0.03 [b]	6.35 ± 0.03 [c]
	50	9.02 ± 0.04 [a]	8.52 ± 0.01 [b]	8.01 ± 0.02 [c]	6.52 ± 0.03 [d]
	60	10.53 ± 0.02 [a]	9.54 ± 0.04 [b]	8.85 ± 0.03 [c]	6.68 ± 0.05 [d]
Fruit Firmness (N)	0	58.72 ± 0.02 [a]	58.62 ± 0.03 [a]	58.71 ± 0.01 [a]	58.63 ± 0.04 [a]
	10	52.51 ± 0.03 [a]	52.62 ± 0.03 [a]	52.54 ± 0.04 [a]	52.53 ± 0.03 [a]
	20	46.03 ± 0.04 [a]	47.41 ± 0.03 [a]	48.32 ± 0.06 [a]	47.44 ± 0.03 [a]
	30	40.72 ± 0.03 [b]	42.01 ± 0.04 [b]	45.23 ± 0.05 [a]	46.22 ± 0.06 [a]
	40	36.31 ± 0.05 [b]	38.81 ± 0.06 [b]	43.12 ± 0.03 [a]	44.13 ± 0.02 [a]
	50	30.92 ± 0.04 [c]	31.43 ± 0.03 [c]	37.92 ± 0.02 [b]	42.01 ± 0.03 [a]
	60	25.62 ± 0.06 [d]	29.31 ± 0.02 [c]	35.91 ± 0.05 [b]	41.82 ± 0.04 [a]
Titratable Acidity (%)	0	0.42 ± 0.01 [a]	0.42 ± 0.02 [a]	0.42 ± 0.01 [a]	0.42 ± 0.01 [a]
	10	0.36 ± 0.02 [a]	0.37 ± 0.00 [a]	0.38 ± 0.01 [a]	0.36 ± 0.01 [a]
	20	0.28 ± 0.00 [a]	0.29 ± 0.02 [a]	0.36 ± 0.03 [a]	0.34 ± 0.01 [a]
	30	0.18 ± 0.01 [d]	0.26 ± 0.02 [b]	0.25 ± 0.03 [c]	0.27 ± 0.02 [a]
	40	0.12 ± 0.02 [d]	0.20 ± 0.02 [c]	0.22 ± 0.03 [b]	0.21 ± 0.01 [a]
	50	0.09 ± 0.01 [d]	0.15 ± 0.02 [c]	0.18 ± 0.00 [b]	0.16 ± 0.01 [a]
	60	0.06 ± 0.00 [d]	0.12 ± 0.01 [c]	0.15 ± 0.02 [b]	0.14 ± 0.03 [a]
Total Soluble Solids (%)	0	15.12 ±0.16 [a]	15.12 ± 0.12 [a]	15.12 ± 0.20 [a]	15.12 ± 0.11 [a]
	10	15.53 ± 0.14 [a]	15.41 ± 0.12 [a]	15.62 ± 0.18 [a]	15.85 ± 0.19 [a]
	20	15.83 ± 0.15 [a]	15.65 ± 0.16 [a]	15.85 ± 0.12 [a]	15.92 ± 0.11 [a]
	30	12.23 ± 0.10 [b]	13.97 ± 0.11 [a]	14.47 ± 0.14 [a]	14.32 ± 0.10 [a]
	40	11.36 ± 0.15 [c]	13.05 ± 0.22 [b]	14.32 ± 0.10 [ab]	13.85 ± 0.08 [a]
	50	10.85 ± 0.12 [c]	12.51 ± 0.11 [b]	13.22 ± 0.10 [ab]	13.55 ± 0.21 [a]
	60	9.81 ± 0.19 [c]	11.32 ± 0.22 [b]	12.23 ± 0.08 [b]	13.20 ± 0.10 [a]

* Means within each row with the different letters indicate significant difference ($p < 0.05$) between treatments.

3.2.3. Respiratory Rate and Ethylene Production

As shown in Figure 4A, the CK and LFT treatment groups reached their first respiratory peak on the 20th day with 32.96 and 30 mg CO_2 kg^{-1} h^{-1}, respectively, while the LFT + O_3 and LFT + TiO_2 groups had only one respiratory peak on the 50th day of 24.68 and 26.52 mg CO_2 kg^{-1} h^{-1}. The CK and LFT treatment groups reached their maximum value of 35.25 and 32.25 mg CO_2 kg^{-1} h^{-1}, respectively, on the 50th day. In addition, the respiratory rate of LFT + O_3 treated peaches was significantly lower than the other three groups from days 20 to 50 ($p < 0.05$). Ozone mainly inhibits the respiration by inhibiting oxidative phosphorylation of mitochondria of fruit cells and the normal electron-transport respiratory chain [45]. However, although the treatments of LFT + O_3 and LFT + TiO_2 were lower than the CK and LFT groups, there was no significant difference between them ($p > 0.05$). As a typical climacteric fruit, inhibiting or delaying the emergence of respiratory peak is the key measure to maintaining the storage quality of peach fruit [46]. It indicated that the combination of LFT and a TiO_2 photocatalysis reactor or ozone could significantly inhibit the peach's breathing. Similar results were reported by Han et al. [34] for ozone treatment on black mulberry, and Tao et al. [47] for a chitosan/nano-TiO_2 composite film treatment on pears.

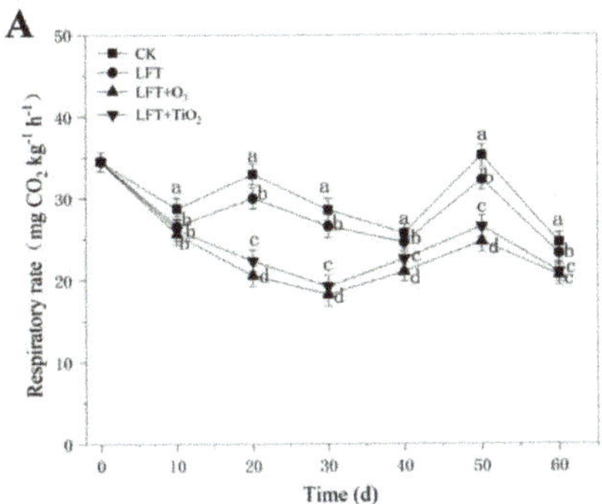
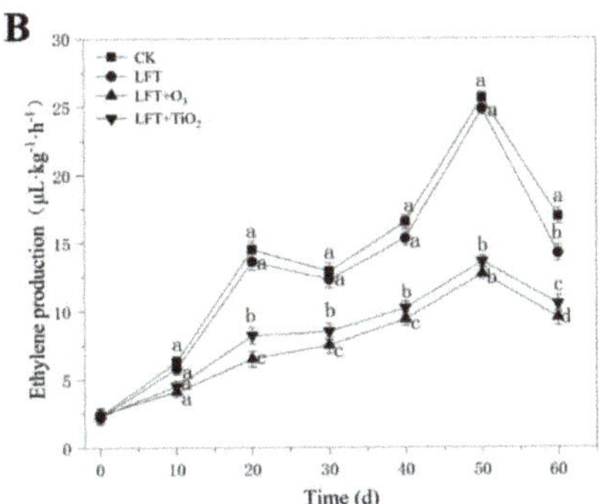

Figure 4. Effect of control (CK), LFT, LFT combined with ozone treatment (LFT + O_3) and LFT combined with TiO_2 photocatalytic (LFT + TiO_2) treatments on the respiratory rate and ethylene production of peaches during storage at 0 °C for 60 d. Values are expressed as means ± SD (n = 10). Different letters (a–d) indicate significant differences among treatments for each sampling time at $p < 0.05$. (**A**) Respiratory rate. (**B**) Ethylene production.

The release of ethylene could accelerate the ripening and senescence of peaches during storage [48]. As shown in Figure 4B, the maximum ethylene release of the LFT + O_3 and LFT + TiO_2 groups was 12.7 and 13.6 μL kg^{-1} h^{-1}, respectively, which were significantly lower than of those of the CK and LFT groups at 40 d storage ($p < 0.05$). The difference in ethylene production between the CK group and the LFT + O_3 group fruit was highly significant, and CK group fruit provided more than twice the ethylene production than that of LFT + O_3 group fruit. Hawkins et al. [49] demonstrated that ozone and its anions have certain effects on endogenous ethylene degradation. In addition, our results show that LFT combined with ozone and TiO_2 photocatalysis have a good synergistic effect in degrading ethylene during peach storage. Similar results have been reported in a previous study where the presence of TiO_2 and UV-A light can remove ethylene gas from the storage atmosphere [23]. However, although the ethylene production of LFT treatment was also lower than that of the CK group, the differences were not significant ($p > 0.05$), which indicates that LFT has little effect on regulating the release of ethylene.

3.2.4. PPO Activity and Total Phenolics Content

PPO has long been considered to be a major factor leading to fruit discoloration after harvest. As shown in Figure 5A, there was no significant difference in PPO activities among all treatments during the first 20 d of storage. However, PPO activity declined rapidly in both LFT treated and CK group fruit starting at 40 d, which might be because the fruit became senescent and over-ripe. Greater efficacies in inhibiting the PPO activities were found in both LFT + O_3 and LFT + TiO_2 treated fruit than that of CK group fruit after 20 d. Furthermore, at the 60th day of storage, reduction of PPO activity can be achieved by the LFT + TiO_2 treatment, compared with the other treatments.

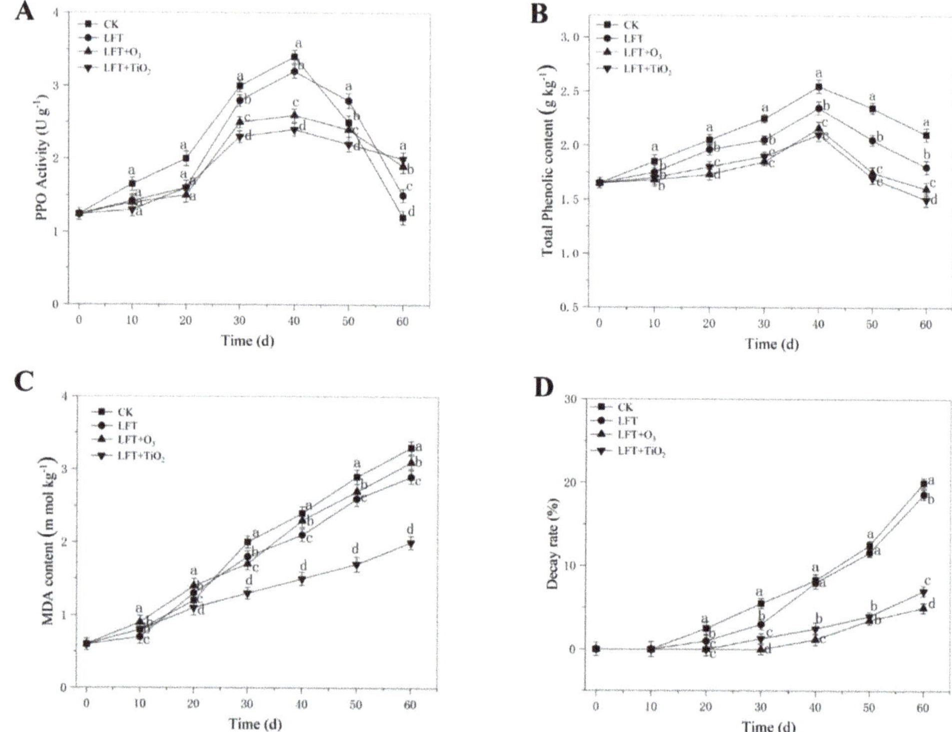

Figure 5. Effect of control (CK), LFT, LFT combined with ozone treatment (LFT + O$_3$) and LFT combined with TiO$_2$ photocatalytic (LFT + TiO$_2$) treatments on the PPO activity, total phenolics content, decay rate, and MDA content of peaches fruit during storage at 0 °C for 60 d. Values are expressed as means ± SD (n = 10). Different letters (a–d) indicate significant differences among treatments for each sampling time at $p < 0.05$. (**A**) PPO activity. (**B**) Total phenolics content. (**C**) Malondialdehyde content. (**D**) Decay rate.

The total phenol content progressively increased for the first 40 d and then decreased during storage in all the treatments of peach fruit, and the fruit treated with LFT + TiO$_2$ possessed the lowest total phenol content and the slowest increase rate (Figure 5B). This may be related to the TiO$_2$ photocatalysis. A similar result was reported by Li et al. [42] that nano-TiO$_2$–low-density polyethylene packaging could inhibit the biosynthesis of phenolics. As the substrate of enzymatic browning, the total phenol content exhibited positive correlated responses in the degree of browning [50]. Reducing the content of total phenol and PPO activity is one of the main modes of fruit resistance to browning [51]. Here, the LFT+O$_3$ and LFT + TiO$_2$ treatments can effectively reduce the formation of total phenols and inhibit the activity of PPO. In our study, higher PPO activity and total phenol content were found in LFT + O$_3$ group fruit after 40 d of storage. It might be related to the physiological damage of the peach fruit being excessively exposed to ozone. The report of Ong et al. [52] showed that the balance between oxidative and reductive processes might be destroyed due to repeated ozone treatments, which then promotes the oxidation of phenolics, resulting in browning. Overall, LFT + TiO$_2$ treatment could effectively reduce the PPO activity and the corresponding total phenolic content in peaches.

3.2.5. Malondialdehyde (MDA) Content

As shown in Figure 5C, MDA content increased substantially in all treatments. However, the MDA content of peach fruit in the LFT + TiO$_2$ group was remarkably restricted: only 60.6% of the decay rate of the CK group fruit at the end of storage time. LFT treatment reduced and delayed the accumulation of MDA, and MDA content in LFT treated fruit was 2.9 mmol kg^{-1} on day 60, showing about 12% less than that of the CK fruit. The cell membrane changes from a gel phase to a liquid crystal phase at large temperature fluctuations, which increases the risk of semi-permeable membrane loss [53]. However, higher MDA content of fruit was observed in the LFT+O$_3$ treatment after 40 d, which may be due to the lack of free radical scavenging ability of ozone-treated fruits at low temperature [54]. In addition, the interaction of phenolic compounds with PPO is enhanced following damage of membrane integrity, which leads to tissue deterioration or senescence of the fruit [55]. In this study, it was observed that the LFT + TiO$_2$ treatment reduced the MDA content. Our result is in agreement with a previous study, which reported that nano-TiO$_2$ films can decrease the accumulation of MDA content of Ginkgo biloba seeds [56].

3.2.6. Decay Rate

As shown in Figure 5D, fruits of the CK and LFT groups decayed during the first 20 d, and the degree of decay was significantly higher than of those of the LFT + O$_3$ and LFT + TiO$_2$ groups. However, the decay symptoms were observed after 30 d in fruits of LFT + TiO$_2$ and LFT + O$_3$ groups, and the decay rate of peach fruit was significantly reduced. The decay rates of LFT + TiO$_2$ and LFT + O$_3$ treatments were 65.2% and 75.3% lower than that of the control group at 60 d, respectively. Hoffmann et al. [57] reported that the superoxide anion radicals ($\cdot$O$_2^-$) and hydroxyl radicals ($\cdot$OH) produced by the TiO$_2$ photocatalytic reactor under the irradiation of light with a specific wavelength have strong oxidative decomposition capability to kill bacteria by damaging the proteins in the cell membrane. Among all treatments, the LFT + O$_3$ treatment employed in this study resulted in a significant effect on the decay rate of peach fruit, which might be related to the effective inhibition of ozone on microorganisms. Victorin [58] reported that ozone could destroy microorganisms by oxidizing cellular components such as sulfhydryl groups in amino acids and enzymes in cell membranes. Similarly, ozone could reduce the decay rate of blackberries after harvest [59].

3.2.7. Scanning Electron Microscopy (SEM) Observation

In this study, the micromorphology of the peel and flesh structure of peaches at the end of storage was observed. As shown in Figure 6A, significant destruction, folding, and deformation of flesh tissue were observed in the fruit of both the CK and the LFT groups at the end of storage. In contrast, LFT combined with ozone or a TiO$_2$ photocatalytic reactor maintained an integrated and uniform tissue structure of peach fruit. This may reveal the ability of ozone and TiO$_2$ photocatalysis to maintain normal physiological metabolism of peach fruit and inhibit microbial reproduction. SEM images of a control fruit surface-section showed deformed stomata, indicating the loss of moisture control function in epidermal cells (Figure 6B1). Conversely, the stomata were regular on the fruit surface of the LFT + TiO$_2$ group, the morphology of guard cells was complete, the closed status could effectively suppress water loss (Figure 6B4). However, severe hair loss and stomatal closure occurred in the LFT + O$_3$ treatment group due to repeated ozone fumigation. Similar phenomenon was also reported by Han et al. [34] where the stomas of black mulberry peel was closed by ozone treatment. In conclusion, the LFT + TiO$_2$ photocatalytic treatment significantly suppressed the degradation of flesh and epidermal tissue of peach fruit, allowing the maintenance of their morphological features.

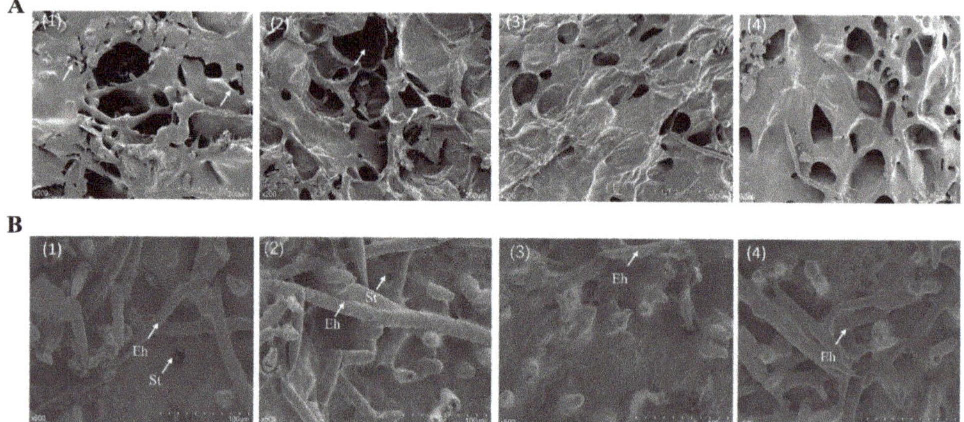

Figure 6. SEM images of different treatments on the flesh structure (**A**) and peel (**B**) of peach fruit on the 60th day. (**1**) CK; (**2**) LFT; (**3**) LFT + O$_3$; (**4**) LFT + TiO$_2$. The stomata and epidermis of peach skin are represented by St and Eh, respectively.

4. Conclusions

In the present study, low-temperature fluctuations combined with either ozone fumigation or a TiO$_2$ photocatalysis reactor could effectively reduce the decay and respiration rates, and degrade ethylene during refrigerated storage. However, slight oxidation and damage of peach fruit were found in the LFT + O$_3$ treatment during the later stage of storage. In addition, the LFT + TiO$_2$ treatment was superior to LFT + O$_3$ in maintaining fruit color and microstructure, inhibiting the enzyme activity of PPO, preventing the substrate generation of total phenol, and extending the shelf-life of peach fruit. In summary, the LFT + TiO$_2$ treatment provided a more appropriate air composition for peach storage, which was conducive to prolonging the postharvest life and ensuring the quality of fruit during storage.

Author Contributions: Designed the experiments, X.L., J.L., X.J., and Z.Z.; Wrote the manuscript, X.J.; Performed the experiments and analyzed the data, X.J., M.D., Y.Z., and L.C.; Contributed to the language editing, J.S. and W.Y. All authors have read and agreed to the published version of the manuscript.

Funding: This research was funded by the support of the Sci-Tech Project of Xinjiang Production and Construction Corps of China (2019AB024), National Key Technology Research and Development Program of the Ministry of Science and Technology of China (2017YFD0401305,2018YFD0401305), Natural Science Foundation of Ningxia (2019AAC03285), Natural Science Foundation of Tianjin (No:18JCQNJC14600), Tianjin Sci-Tech Project (18ZXRHNC00070,19YFZCSN00460).

Acknowledgments: The authors thank Baoshuang Tian (from the Tianjin Gasin-DH Preservation Technology Co., Ltd., China) for the technical assistance in the experimental device of TiO$_2$ photocatalytic.

Conflicts of Interest: The authors declare no conflict of interest.

References

1. Wang, J.; You, Y.; Chen, W.; Xu, Q.; Wang, J.; Liu, Y.; Song, L.; Wu, J. Optimal hypobaric treatment delays ripening of honey peach fruit via increasing endogenous energy status and enhancing antioxidant defence systems during storage. *Postharvest Biol. Technol.* **2015**, *101*, 1–9. [CrossRef]
2. Hussain, P.R.; Meena, R.S.; Dar, M.A.; Wani, A.M. Studies on enhancing the keeping quality of peach (Prunus persica Bausch) Cv. Elberta by gamma-irradiation. *Radiat. Phys. Chem.* **2008**, *77*, 473–481. [CrossRef]
3. Budde, C.O.; Polenta, G.; Lucangeli, C.D.; Murray, R.E. Air and immersion heat treatments affect ethylene production and organoleptic quality of 'Dixiland' peaches. *Postharvest Biol. Technol.* **2006**, *41*, 32–37. [CrossRef]

4. Zhou, D.; Sun, Y.; Li, M.; Zhu, T.; Tu, K. Postharvest hot air and UV-C treatments enhance aroma-related volatiles by simulating the lipoxygenase pathway in peaches during cold storage. *Food Chem.* **2019**, *292*, 294–303. [CrossRef]

5. Santana, L.R.; Benedetti, B.; Sigrist, J. Sensory characteristics of 'Douradão' peaches submitted to modified atmosphere packaging. *Rev. Bras. Frutic.* **2010**, *32*, 700–708. [CrossRef]

6. Artés, F.; Cano, A.; Fernández-Trujillo, J.P. Pectolytic Enzyme Activity during Intermittent Warming Storage of Peaches. *J. Food Sci.* **1996**, *61*, 311–314. [CrossRef]

7. Aung, M.M.; Chang, Y.S. Temperature management for the quality assurance of a perishable food supply chain. *Food Control* **2014**, *40*, 198–207. [CrossRef]

8. Urquiola, A.; Alvarez, G.; Flick, D. Frost formation modeling during the storage of frozen vegetables exposed to temperature fluctuations. *J. Food Eng.* **2017**, *214*, 16–28. [CrossRef]

9. Ho, S.H.; Rosario, L.; Rahman, M.M. Numerical simulation of temperature and velocity in a refrigerated warehouse. *Int. J. Refrig.* **2010**, *33*, 1015–1025. [CrossRef]

10. Brummell, D.A.; Dal Cin, V.; Lurie, S.; Crisosto, C.H.; Labavitch, J.M. Cell wall metabolism during the development of chilling injury in cold-stored peach fruit: Association of mealiness with arrested disassembly of cell wall pectins. *J. Exp. Bot.* **2004**, *55*, 2041–2052. [CrossRef]

11. Lyons, J.M. Chilling injury in plants. *Ann. Rev. Plant Phys.* **1973**, *20*, 423–446. [CrossRef]

12. Talasila, P.C.; Chau, K.V.; Brecht, J.K. Effects of gas concentrations and temperature on O_2 consumption of strawberries. *Trans. ASAE* **1992**, *35*, 221–224. [CrossRef]

13. Lehmann, D.C.; Ferguson, J.E. A modified jacketed cold storage design. *Int. J. Refrig.* **1984**, *7*, 186–189. [CrossRef]

14. Peters, J.A.; McLane, D.T. Storage life of pink shrimp held in commercial and jacketed cold-storage rooms. *Commer. Fish. Rev.* **1959**, *21*, 1–7.

15. Lentz, C.P.; Rooke, E.A. Use of the jacketed room system for cool storage. *Food Technol.* **1957**, *11*, 257–259.

16. Lentz, C.P.; Van den Berg, L.; Jorgensen, E.G.; Sawler, R. The design and operation of a jacketed vegetable storage. *Can. Inst. Food Sci. Technol. J.* **1971**, *4*, 19–23. [CrossRef]

17. Raghavan, G.S.V.; Bovell, R.; Chayet, M. Storability of fresh carrots in a simulated jacketed storage. *Trans. Am. Soc. Agric. Eng.* **1980**, *23*, 1521–1524. [CrossRef]

18. Khadre, M.A.; Yousef, A.E.; Kim, J.G. Microbiological Aspects of Ozone Applications in Food: A Review. *J. Food Sci.* **2001**, *66*, 1242–1252. [CrossRef]

19. Bodaghi, H.; Mostofi, Y.; Oromiehie, A.; Zamani, Z.; Ghanbarzadeh, B.; Costa, C.; Conte, A.; Del Nobile, M.A. Evaluation of the photocatalytic antimicrobial effects of a TiO_2 nanocomposite food packaging film by in vitro and in vivo tests. *LWT Food Sci. Technol.* **2013**, *50*, 702–706. [CrossRef]

20. Cho, M.; Chung, H.; Choi, W.; Yoon, J. Linear correlation between inactivation of E. coli and OH radical concentration in TiO2 photocatalytic disinfection. *Water Res.* **2004**, *38*, 1069–1077. [CrossRef]

21. Huang, Z.; Maness, P.-C.; Blake, D.M.; Wolfrum, E.J.; Smolinski, S.L.; Jacoby, W.A. Bactericidal mode of titanium dioxide photocatalysis. *J. Photochem. Photobiol. A* **2000**, *130*, 163–170. [CrossRef]

22. Shahbaz, H.M.; Kim, S.; Hong, J.; Kim, J.U.; Lee, D.U.; Ghafoor, K.; Park, J. Effects of TiO_2–UVC photocatalysis and thermal pasteurisation on microbial inactivation and quality characteristics of the Korean rice-and-malt drink sikhye. *Int. J. Food Sci. Tech.* **2016**, *51*, 123–132. [CrossRef]

23. Maneerat, C.; Hayata, Y. Efficiency of TiO_2 photocatalytic reaction on delay of fruit ripening and removal of off-flavors from the fruit storage atmosphere. *Trans. ASAE* **2006**, *49*, 833–837. [CrossRef]

24. Miller, F.A.; Silva, C.L.; Brandão, T.R. A review on ozone-based treatments for fruit and vegetables preservation. *Food Eng. Rev.* **2013**, *5*, 77–106. [CrossRef]

25. Kying, M.; Ali, A.; Alderson, P.; Forney, C.; Malaysia, S.D.E.; Tunku, A.; Rahman, J.; Barat Kampar, P. Effect of different concentrations of ozone on physiological changes associated to gas exchange, fruit ripening, fruit surface quality and defence-related enzymes levels in papaya fruit during ambient storage. *Sci. Hortic.* **2014**, *179*, 163–169.

26. Santana, L.R.R.D.; Benedetti, B.C.; Sigrist, J.M.M.; Sato, H.H.; Anjos, V.D.D.A. Effect of controlled atmosphere on postharvest quality of 'Douradão' peaches. *Food Sci. Technol.* **2011**, *31*, 231–237. [CrossRef]

27. Liu, X.; Lu, Y.; Yang, Q.; Yang, H.; Li, Y.; Zhou, B.; Li, T.; Gao, Y.; Qiao, L. Cod peptides inhibit browning in fresh-cut potato slices: A potential anti-browning agent of random peptides for regulating food properties. *Postharvest Biol. Technol.* **2018**, *146*, 36–42. [CrossRef]

28. Wang, Q.; Cao, Y.; Zhou, L.; Jiang, C.-Z.; Feng, Y.; Wei, S. Effects of postharvest curing treatment on flesh colour and phenolic metabolism in fresh-cut potato products. *Food Chem.* **2015**, *169*, 246–254. [CrossRef]

29. Zhang, Y.; Gong, Y.; Chen, L.; Peng, Y.; Wang, Q.; Shi, J. Hypotaurine delays senescence of peach fruit by regulating reactive oxygen species metabolism. *Sci. Hortic.* **2019**, *253*, 295–302. [CrossRef]

30. Zhong, Q.; Xia, W. Effect of 1-methylcyclopropene and/or chitosan coating treatments on storage life and quality maintenance of Indian jujube fruit. *LWT Food Sci. Technol.* **2007**, *40*, 404–411.

31. Li, X.; Li, W.; Jiang, Y.; Ding, Y.; Yun, J.; Tang, Y.; Zhang, P. Effect of nano-ZnO-coated active packaging on quality of fresh-cut 'Fuji' apple. *Int. J. Food Sci. Technol.* **2011**, *46*, 1947–1955. [CrossRef]

32. Yang, Z.; Cao, S.; Su, X.; Jiang, Y. Respiratory activity and mitochondrial membrane associated with fruit senescence in postharvest peaches in response to UV-C treatment. *Food Chem.* **2014**, *161*, 16–21. [CrossRef]

33. Huan, C.; An, X.J.; Yu, M.L.; Jiang Li, J.; Ma, R.J.; Tu, M.G.; Yu, Z.F. Effect of combined heat and 1-MCP treatment on the quality and antioxidant level of peach fruit during storage. *Postharvest Biol. Technol.* **2018**, *145*, 193–202. [CrossRef]

34. Han, Q.; Gao, H.; Chen, H.; Fang, X.; Wu, W. Precooling and ozone treatments affects postharvest quality of black mulberry (Morus nigra) fruit. *Food Chem.* **2017**, *221*, 1947–1953. [CrossRef] [PubMed]

35. Piccolella, S.; Fiorentino, A.; Pacifico, S.; D'Abrosca, B.; Uzzo, P.; Monaco, P. Antioxidant Properties of Sour Cherries (Prunus cerasus L.): Role of Colorless Phytochemicals from the Methanolic Extract of Ripe Fruit. *J. Agric. Food Chem.* **2008**, *56*, 1928–1935. [CrossRef]

36. Dhindsa, R.S.; Plumbdhindsa, P.; Thorpe, T.A. Leaf senescence: Correlated with increased levels of membrane permeability and lipid peroxidation, and decreased levels of superoxide dismutase and catalase. *J. Exp. Bot.* **1981**, *32*, 93–101. [CrossRef]

37. Nicola, B.M.; Verlinden, B.; Beuselinck, A.; Jancsok, P.; Valéry, Q.; Scheerlinck, N.; Verboven, P.; De Baerdemaeker, J. Propagation of stochastic temperature fluctuations in refrigerated fruit: Propagation des fluctuations de température stochastiques dans des fruit réfrigérés. *Int. J. Refrig.* **1999**, *22*, 81–90. [CrossRef]

38. Cameron, A.C.; Talasila, P.C.; Joles, D.J. Predicting the film permeability needs for modified-atmosphere packaging of lightly processed fruit and vegetables. *HortScience* **1995**, *30*, 25–34. [CrossRef]

39. Forney, C.F.; Song, J.; Hildebrand, P.D.; Fan, L.; McRae, K.B. Interactive effects of ozone and 1-methylcyclopropene on decay resistance and quality of stored carrots. *Postharvest Biol. Technol.* **2007**, *45*, 341–348. [CrossRef]

40. Bridges, D.F.; Rane, B.; Wu, V.C.H. The effectiveness of closed-circulation gaseous chlorine dioxide or ozone treatment against bacterial pathogens on produce. *Food Control* **2018**, *91*, 261–267. [CrossRef]

41. Ali, A.; Ong, M.K.; Forney, C.F. Effect of ozone pre-conditioning on quality and antioxidant capacity of papaya fruit during ambient storage. *Food Chem.* **2014**, *142*, 19–26. [CrossRef] [PubMed]

42. Li, D.; Ye, Q.; Jiang, L.; Luo, Z. Effects of nano-TiO_2 -LDPE packaging on postharvest quality and antioxidant capacity of strawberry (Fragaria ananassa Duch.) stored at refrigeration temperature. *J. Sci. Food Agric.* **2016**, *97*, 1116–1123. [CrossRef] [PubMed]

43. Ge, Y.; Chen, Y.; Li, C.; Wei, M.; Li, X.; Li, S.; Lu, S.; Li, J. Effect of trisodium phosphate dipping treatment on the quality and energy metabolism of apples. *Food Chem.* **2019**, *274*, 324–329. [CrossRef] [PubMed]

44. Xing, Y.; Yang, H.; Guo, X.; Bi, X.; Liu, X.; Xu, Q.; Wang, Q.; Li, W.; Li, X.; Shui, Y.; et al. Effect of chitosan/Nano-TiO2 composite coatings on the postharvest quality and physicochemical characteristics of mango fruits. *Sci. Hortic.* **2020**, *263*, 109–135. [CrossRef]

45. Lee, T.T. Inhibition of oxidative phosphorylation and respiration by ozone in Tobacco Mitochondria. *Plant Physiol.* **1967**, *42*, 691–696. [CrossRef]

46. Chun, J.P.; Seo, J.S.; Kim, M.S.; Lim, B.S.; Ahn, Y.J.; Hwang, Y.S. Effects of 1-MCP and storage condition on shelf life and quality of 'Janghowon Hwangdo' peach (Prunus persica Batsch). *Kor. J. Hort. Sci. Technol.* **2010**, *28*, 585–592.

47. Tao, X.; Wang, M. Study on the Optimal Composition of Chitosan/Nano- TiO_2 Composite Film and Its Application in the Preservation of Jinqiu Pear. *Agric. Biotechnol.* **2015**, *4*, 20–23.

48. Huang, D.; Hu, S.; Zhu, S.; Feng, J. Regulation by nitric oxide on mitochondrial permeability transition of peaches during storage. *Plant Physiol. Biochem.* **2019**, *138*, 17–25. [CrossRef]

49. Hawkins, M.; Kohlmiller, C.K.; Andrews, L. Matrix infrared spectra and photolysis and pyrolysis of isotopic secondary ozonides of ethylene. *J. Phys. Chem.* **1982**, *86*, 3154–3166. [CrossRef]

50. Coseteng, M.Y.; Lee, C.Y. Changes in apple polyphenoloxidase and polyphenol concentrations in relation to degree of browning. *J. Food Sci.* **2010**, *52*, 985–989. [CrossRef]

51. Gao, H.; Chai, H.; Cheng, N.; Cao, W. Effects of 24-epibrassinolide on enzymatic browning and antioxidant activity of fresh-cut lotus root slices. *Food Chem.* **2017**, *217*, 45–51. [CrossRef] [PubMed]

52. Ong, M.K.; Ali, A.; Alderson, P.G.; Forney, C.F. Effect of different concentrations of ozone on physiological changes associated to gas exchange, fruit ripening, fruit surface quality and defence-related enzymes levels in papaya fruit during ambient storage. *Sci. Hortic.* **2014**, *179*, 163–169. [CrossRef]

53. Jin, P.; Zhu, H.; Wang, L.; Shan, T.; Zheng, Y. Oxalic acid alleviates chilling injury in peach fruit by regulating energy metabolism and fatty acid contents. *Food Chem.* **2014**, *161*, 87–93. [CrossRef] [PubMed]

54. Calatayud, A.; Barreno, E. Chlorophyll a fluorescence, antioxidant enzymes and lipid peroxidation in tomato in response to ozone and benomyl. *Environ. Pollut.* **2001**, *115*, 283–289. [CrossRef]

55. Lurie, S.; Crisosto, C.H. Chilling injury in peach and nectarine. *Postharvest Biol. Technol.* **2005**, *37*, 195–208. [CrossRef]

56. Tian, F.; Chen, W.; Wu, C.E.; Kou, X.; Fan, G.; Li, T.; Wu, Z. Preservation of Ginkgo biloba seeds by coating with chitosan/nano-TiO2 and chitosan/nano-SiO2 films. *Int. J. Biol. Macromol.* **2019**, *126*, 917–925. [CrossRef]

57. Hoffmann, M.R.; Martin, S.T.; Choi, W.; Bahnemann, D.W. Environmental applications of semiconductor photocatalysis. *Chem. Rev.* **1995**, *95*, 69–96. [CrossRef]

58. Victorin, K. Review of genotoxicity of ozone. *Mutat. Res.* **1992**, *277*, 221–238. [CrossRef]

59. Barth, M.; Zhou, M.; Mercier, C.; Payne, J. Ozone storage effects on anthocyanin content and fungal growth in blackberries. *J. Food Sci.* **1994**, *60*, 1286–1288. [CrossRef]

© 2020 by the authors. Licensee MDPI, Basel, Switzerland. This article is an open access article distributed under the terms and conditions of the Creative Commons Attribution (CC BY) license (http://creativecommons.org/licenses/by/4.0/).

Review

Botrytis cinerea and Table Grapes: A Review of the Main Physical, Chemical, and Bio-Based Control Treatments in Post-Harvest

Nicola De Simone [1], Bernardo Pace [2], Francesco Grieco [3], Michela Chimienti [4], Viwe Tyibilika [5], Vincenzo Santoro [6], Vittorio Capozzi [2,*], Giancarlo Colelli [1], Giuseppe Spano [1] and Pasquale Russo [1]

[1] Department of the Sciences of Agriculture, Food and Environment, University of Foggia, Via Napoli 25, 71122 Foggia, Italy; nicola_desimone.552001@unifg.it (N.D.S.); giancarlo.colelli@unifg.it (G.C.); giuseppe.spano@unifg.it (G.S.); pasquale.russo@unifg.it (P.R.)

[2] Institute of Sciences of Food Production, National Research Council of Italy (CNR), c/o CS-DAT, Via Michele Protano, 71121 Foggia, Italy; bernardo.pace@ispa.cnr.it

[3] Institute of Sciences of Food Production, National Research Council of Italy (CNR), Via Prov.le Lecce-Monteroni, 73100 Lecce, Italy; francesco.grieco@ispa.cnr.it

[4] InResLab Scarl, Contrada Baione, 70043 Monopoli, Italy; m.chimienti@inreslab.org

[5] AgroSup Dijon, 21079 Dijon CEDEX, France; viwetyibilika@gmail.com

[6] A.B.A. Mediterranea s.c.a.r.l., Via Parini, 1, 74013 Ginosa, Italy; enzo.santoro@abamediterranea.it

* Correspondence: vittorio.capozzi@ispa.cnr.it; Tel.: +39-0881-630201

Received: 23 June 2020; Accepted: 11 August 2020; Published: 19 August 2020

Abstract: Consumers highly appreciate table grapes for their pleasant sensory attributes and as good sources of nutritional and functional compounds. This explains the rising market and global interest in this product. Along with other fruits and vegetables, table grapes are considerably perishable post-harvest due to the growth of undesired microorganisms. Among the microbial spoilers, *Botrytis cinerea* represents a model organism because of its degrading potential and the huge economic losses caused by its infection. The present review provides an overview of the recent primary physical, chemical, and biological control treatments adopted against the development of *B. cinerea* in table grapes to extend shelf life. These treatments preserve product quality and safety. This article also focuses on the compliance of different approaches with organic and sustainable production processes. Tailored approaches include those that rely on controlled atmosphere and the application of edible coating and packaging, as well as microbial-based activities. These strategies, applied alone or in combination, are among the most promising solutions in order to prolong table grape quality during cold storage. In general, the innovative design of applications dealing with hurdle technologies holds great promise for future improvements.

Keywords: table grapes; *Botrytis cinerea*; grey mould; spoilage microbes; post-harvest; modified atmosphere packaging (MAP); ozone (O_3); antimicrobial compounds; preservatives; biocontrol

---

1. Introduction

Viticulture is one of the major forms of fruit crop cultivation worldwide, and its global diffusion contributes considerably to human nutrition. The fruit has a non-climacteric character with a quite low rate of physiological activity. Grapes (*Vitis vinifera* L.) are essential not only for wine production but also for fresh consumption. Table grapes are highly appreciated by consumers, primarily because of their sensory attributes, but also because of their vitamins and bioactive compounds (e.g., flavonoids) [1]. More than 27 million tons of table grapes are produced worldwide annually (an increase of 71% since 2000), and about 4.2 million tons were exported among countries in 2014 [2]. Accordingly, increasing attention has been paid to lengthening the shelf-life of table grapes for export. Prolonged storage time preserves marketability and adds value; however, it is often associated with a decrease in overall product quality. In general, several factors, including bunch dehydration, rachis browning, peel colour changes, lacerations and colonization by various spoilage fungi result in significant economic losses.

Among other factors, fungal decay represents the principal factor responsible for post-harvest deterioration in table grapes [3]. *Botrytis cinerea* is the main biological cause of post-harvest problems since it is accountable for grey mould formation [4]. Indeed, this undesired fungus is ranked second in the "world top 10 fungal pathogens in molecular plant pathology" in terms of economic and scientific relevance, preceded only by *Magnaporthe oryzae* [5]. Fungal spores are generally present on the surface of fruits, and, during post-harvest handling the berries can supply a suitable environment for spore germination (mainly the damaged fruits) (Figure 1).

Figure 1. Effect of grey mould on cold-stored cv. "Italia" table grape berries. Image from Ahmed et al. [3].

Moreover, the infection can occur during storage, marketing, and even after customer purchase. In the vineyard, high relative air humidity and low environmental temperatures reduce the host's defences. This environment favours the rapid spread of contamination from a single berry to the whole bunch [6,7]. During post-harvest treatments of fruits and vegetables, processing technologies and biotechnologies provide physical, chemical, and biological hurdles to limit the development of undesired microorganisms [8]. Changes in technical and technological solutions, consumer needs, and regulatory framework lead to a continuous evolution of the handling procedures to limit decay induced by spoilage fungi. All of these advances are generally tailored to reducing and averting spoilage growth, but they are more broadly oriented towards optimization of global quality of production, including safety, health properties, and sensory acceptability [9–12].

Among the economic and social trends, attention to sustainable viticulture and organic production represents a field of high interest, as evidenced by the rising number of cultivated hectares worldwide (Figure 2).

Nowadays, this kind of table grape cultivation is still increasing in diffusion and economic importance [13]. The production of organic grapes necessitates compliance with specific regulations that limit the chemicals allowed during production and distribution [14]. In general, organic-labelled products are defined as those from plantations that respect and exploit biodiversity, organic turnovers, and soil structure [14]. The European Union has led the cultivation of organic grapes globally, followed by China, the United States of America, and Turkey [15]. Within Europe, the countries with the most extensive acreages dedicated to organic farming are Spain and Italy (1.9 and 1.4 million hectares, respectively; both contributing more than 100,000 hectares to the increase in organic land observed in Europe) [15].

In recent years, different strategies have been proposed to control *B. cinerea* in order to improve the management of post-harvest decay in table grapes and to prevent quality losses [16–18]. The present review aims to discuss the more recent investigations conceived to control *B. cinerea* decay in table grapes, including the primary physical, chemical, and biological approaches.

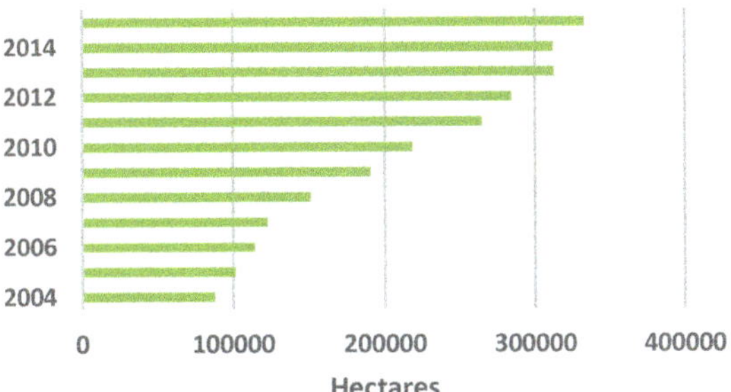

Figure 2. Global area for the cultivation of organic grapes in the period 2004–2015. Source: Research Institute of Organic Agriculture (FiBL) and IFOAM—Organics International—SOEL magazine (2006–2017).

2. Physical Methods to Control *B. cinerea* in Table Grapes

Physical technologies mainly include modification of several parameters such as temperature, absolute and relative gas pressure, UV irradiation, and sonication. Table grapes for fresh consumption often need a long period of storage for commercial purposes such as export and ready-to-eat. They are usually stored in chambers with strictly controlled temperature and humidity. To this aim, cold storage (~0 °C) is the primary method to avoid post-harvest infections without affecting the main physicochemical features of the product [19]. However, *B. cinerea* survives at low temperatures, and any variation of temperature can promote water condensation, thus favouring fungal growth and sporulation [20]. In general, physical methods are often considered eco-friendly and residue-free emerging technologies, widely accepted by consumers. Although these methods have been extensively investigated in different fruit and vegetable products, only a few studies report their employment for the reduction of grey mould in table grapes (Table 1).

Surface sanitation is the main strategy implemented to control microbial contamination of fruits and it can be achieved by using different methods. Among these, dipping in hot water (about 50 °C) is an interesting option to prolong the shelf-life of fruits and vegetables [33,34]. Treatments at 50 °C for 10 min, or at 55 °C for 5 min, are sufficient to reduce the fungal growth, maintaining product quality because it does not alter the grape's organoleptic profile [21,22]. Accordingly, it allows for the marketability of minimally-processed and ready-to-eat table grapes [21,22]. Nonetheless, more studies are requested to improve the processing conditions, i.e., temperature and time of exposure against *B. cinerea* contamination.

Table 1. Main physical methods investigated in the last ten years against grey mould decay in table grapes.

Physical Methods	Treatment Intensity	Cultivar	Effects	Ref.
Hot Water Treatments	Dipping for 5 min at 55 °C	Müşküle and Red Globe	Low decay rate after three weeks of cold storage; sensory evaluation results showed no alteration of flavor and taste	[21]
	Dipping for 10 min at 50 °C	Crimson Seedless	Inhibition the microbial growth during storage without significant changes in texture, titratable acidity, and soluble solids content	[22]
Ultrasound	32 kHz at 20 °C for 10 min	Michele Palieri	Combined with putrescine, the treatment maintained high levels of anthocyanins, total phenolic content, antioxidant capacity, sensory acceptability and reduced decay incidence during storage	[23]
UV-C Irradiation	Two times at 6.0 kJ/m^2 for 1 min at 60 cm	Crimson	Combined with chitosan coating, the treatment increased the resveratrol content, maintained sensorial quality, and reduced fungal decay	[24]
High Pressure	0.15 MPa for 24 h at 20 °C	Italia	Reduction of lesion diameter and decay rate after three days of shelf-life	[25]
Electrolyzed oxidizing water	(250 ppm TRC; pH = 6.3–6.5; ORP = 800–900 mV, 1% NaCl) dipping and daily spray	Thompson seedless	Prevention of infection until seven days; 1% of incidence and 2% of severity were reported after 10 days of shelf-life at 25 °C	[26]
CA	12% O_2 + 12% CO_2	Flame Seedless and Crimson Seedless	Combined with CO_2, the treatment limited decay incidence in both naturally and artificially infected grapes	[27]
	0.3 μL/L O_3	Sultanina	Reduction of fungal decay during 40 days of cold storage; no significant alteration of quality characteristics	[28]
	0.1 - 0.3 μL/L O_3	Crimson Seedless	Reduction of natural incidence of decay by approximately 65% after five–eight weeks of storage.	[29]
MAP	Passive modifications packaging-induced	Vittoria and Red Globe	Reduction of weight losses, rachis and berry decay	[30]
	2% O_2 + 5% CO_2	Scarlotta	Combined with O_3, the treatment was efficient in decay control but caused sensorial quality losses (intense stem browning, off-flavors perception) Combined with CO_2, the treatment controlled the concentration of acetaldehyde, preserved rachis chlorophyll content and skin color; also, cumulative decay incidence was reduced	[31]
	Initial concentration of 10% CO_2	Italia	Decay control during 14 days of cold storage, and three days of shelf life, low acetaldehyde, and ethanol accumulation	[32]

Ultraviolet irradiation (UV) (wavelengths between 10 to 400 nanometers (nm)) and sonication by ultrasound are non-thermal treatments considered simple, reliable, and eco-friendly emerging technologies for lengthening the shelf life of fresh fruits during storage. Ultraviolet irradiation C (UV-C, 10–280 nm) treatment induced a general stimulation of the phenylpropanoid pathway, associated with plant defence mechanisms, leading to an increased resistance to the diseases in artificially inoculated berries [24]. UV-C irradiation is effective, with dosages between 0.125 to 0.5 kJ/m^2 at a fixed distance of 25 cm [35]. In a recent study, harvested 'Crimson' red table grapes were exposed to an increased UV-C intensity (6.0 kJ/m^2), for two illumination periods of 1 min with a specific distance of 60 cm and then maintained at 20 °C for 24 h, followed by cold storage [24]. Regarding ultrasound application, Bal et al. [23] demonstrated the effectiveness of this treatment at 32 kHz, in a distilled water chamber at 20 °C for 10 min. Their study produced encouraging results in preserving grape quality throughout storage for 60 days. A reduction of decay rate was shown and evaluated by scoring the number of contaminated berries, from 2.8 (water-treated control) to 1.5 (ultrasound treated grapes), in an acceptability scale from 1 to 5 points (1 = no decay; 5 = over 20 decayed berries per bunch in a box of 5 kg grapes). It is essential to underline that, in the last two studies, both UV irradiation and sonication are also compared to treatments which combine physical methods with biological compounds, such as chitosan (an antimicrobial linear polysaccharide derived from chitin) and putrescine (biogenic diamine, a class of compound with relevant biological properties), respectively.

Few studies are reported on the use of high hydrostatic pressure and electrolyzed oxidizing water (EOW), especially on table grapes. Romanazzi et al. [25] investigated the efficiency of hyperbaric treatments at 0.15 MPa for 24 h, on artificially inoculated 'Italia' table grapes berries, during simulated shelf-life for three days at 20 °C. A significant reduction of the infected berries (from 49.0 to 30.8 %) and of their lesion diameter (from 8.7 to 7.2 mm) was reported for the treated grapes, when compared to control fruits stored at ambient pressure [25]. Electrolyzed oxidizing water is produced through the controlled electrolysis of sodium chloride solutions. Dipping in EOW [250 ppm total residual chlorine (TRC); pH = 6.3–6.5; ORP = 800–900 mV, 1% NaCl] was adequate to prevent the infection of green table grapes artificially contaminated with *B. cinerea* until one week, showing a decay rate of 2% after ten days of storage at 25 °C [26]. Interestingly, a dipping treatment followed by a daily spray of grapes with EOW prevented the infection until 24 days, showing a daily decay rate of 2% after 26 days of storage at 25 °C [26].

The modification of absolute and relative gas pressure, in association with low temperatures during storage, is an important strategy to enhance the shelf life of fruits and vegetables [36]. The main methods include controlled atmosphere (CA) and modified atmosphere packaging (MAP). CA is defined as an atmosphere different than air, applied to commodities in the storage chamber. MAP involves a change in gas environment in packaged commodities, as a result of respiration (passive MAP) or by the different gas permeability of the packaging (active MAP) [37]. The latter method has received considerable attention because of the possibility of maintaining modifications up to consumption [38–40]. In both CA and MAP approaches, the use of different gas composition (e.g., changes in ratio Oxygen (O_2)/Carbon dioxide (CO_2)) aims to minimize the metabolic activity and oxidative phenomena, thus reducing the physiological decay caused by aerobic microorganisms (e.g., *B. cinerea*) [36,39]. In table grapes, an atmosphere with different gas composition, including high CO_2/low O_2 concentrations [41–43], and the addition of O_3 [42], has the effect of reducing decay. Furthermore, this strategy retards senescence, reduces stem and berry respiration, limits rachis browning, and preserves berry firmness [41–43]. However, CO_2 concentrations >10% reportedly promote off-flavor development, rachis and berries' browning [43]. CA with ozone (O_3) at 0.3 μL/L was assessed as the minimum concentration to significantly inhibit decay development, in artificially contaminated berries, up to seven weeks in cold storage [28,44]. Recently, in similar storage conditions, ozone-CA with 0.1 μL/L in the day and 0.3 μL/L at night, was found to effectively reduce grey mould, even after 68 days, with a maximum disease incidence of 2.1%, comparable to weekly SO_2-fumigated grapes [29]. Passive MAP in micro-perforated polypropylene films, was found to have the highest performance in the decay management of 'Vittoria' and 'Red Globe' table grapes [30]. Cefola and Pace [32] reported best results on 'Italia' table grapes, after 14 days of cold storage and three days of shelf-life, by using MAP with an initial concentration of 10% CO_2, both in terms of sensory quality preservation and decay control. Considering that the use of massive doses of gas in a single pre-storage application can be defined as a sanitation procedure, we refer the discussion to chemical methods following section.

3. Chemical Methods to Control *B. cinerea* in Table Grapes

At present, sulphur dioxide (SO_2) remains the main method that is used to control the microbial spoilage of post-harvest fruit commodities. The employment of SO_2 provides long term storage due to its antioxidant, antibacterial, antifungal and anti-browning properties [19,45]. However, excessive residue levels of SO_2 in berry peels can result in quality deterioration, such as bleached berries, production of off-flavour, or hairline disorder [46,47]. Significant health risks to consumers are also reported due to the emergence of allergies, nausea, respiratory distress and skin rashes [48]. For this reason, the United States Environmental Protection Agency (USEPA) categorized SO_2 as a pesticide, with maximum tolerance in final products of 10 ppm, and, more generally, sulphur dioxide residuals on table grapes are internationally regulated, including in the European Union [49,50]. Its use is also excluded from certified "organic" grapes [16]. Therefore, several chemical alternatives have been proposed to replace SO_2 in the restraint of *B. cinerea* in table grapes (Table 2).

The use of conventional synthetic fungicides is generating increasing concern among consumers due to the potential negative effects on human health [61], soil microbiota [62], and on microorganisms beneficial for food and beverage fermentations [63]. Even if the use of conventional synthetic fungicides is forbidden for organic grapes [14], application is widespread to prevent spoilage mould formation in conventional agriculture [64]. Despite the fact that some studies have focused on the positive action of different combinations of synthetic fungicides or bioactive compounds [51], the occurrence of resistant strains of *B. cinerea* has been reported [65]. The most recently introduced class of synthetic fungicides belongs to the Succinate Dehydrogenase Inhibitors (SDHIs) [66]. In 2012, a novel SDHI, named fluopyram, was registered against *B. cinerea* and it was able to control grey mould infections in table grapes, with efficacy of inhibition in the range 80.1–94.4% [52]. However, high risks of rapid occurrence of resistance without appropriate management has already been underlined in other crops [67]. For this reason, alternative control methods are needed. Among these, resistance induced by elicitors, molecules able to activate defence gene expression and enhance their antimicrobial-related pathways [68], is an attractive alternative because it is associated with minor environmental risk. Acibenzolar-S-methyl is a commercial elicitor able to activate the phenylpropanoid pathway, which leads to the accumulation of lignin, phenolic compounds and flavonoids [68]. In table grapes, it can be used as spray aspersion or dipping solution, both with a significant reduction in terms of decay incidence [53].

Table 2. Main chemical methods investigated in the last ten years against grey mould decay in table grapes.

	Molecules	Treatment	Concentration	Cultivar	Effects	Ref.
Liquid	Pyrimethanil	Wound inoculation	50 mg/L	Crimson Seedless	Combined with resveratrol (1 g/L), the treatment reduced disease incidence and lesion diameter	[51]
	Fluopyram	Spraying	250 µg/mL	Italia	Efficacy against fungicide-resistant fungal strains	[52]
	Acibenzolar-S-methyl	Dipping	1% w/v	Italia and Benitaka	Reduction of grey mould development after one month of cold storage and one week of shelf life, without alteration of the physicochemical quality	[53]
	Ethanol	Dipping	32 %	Scarlotta Seedless	Reduction of berries decay until ten weeks of storage	[54]
	$FeSO_4$, NH_4HCO_3, Na_2SiO_3, $NaHCO_3$ and Na_2CO_3	Dipping or spraying	1% w/v	Benitaka	Decay incidence reduced, no impact on berries quality parameters with minor exceptions which were at an acceptable level	[55]
Gas	Ethanol	Vapour-generating bags	-	Red Globe	Comparable to SO_2 treatments in decay control, the treatment enhanced berry colour, but caused stem browning	[56]
	Chlorine dioxide (ClO_2)	Injection in bag	2.5 mg/5 kg	Kyoho	Reduction of berry decay and rachis browning	[57]
	Nitrous oxide (N_2O)	Fumigation	50 µL/L	Munage	Reduction of lesion diameter and decay incidence	[58]
	Carbon dioxide (CO_2)	Fumigation	20 %	Cardinal	The treatment avoided post-harvest losses in terms of water loss, oxidative damage and disease prevention	[59]
		Fumigation	40%	Flame Seedless and Crimson Seedless	Combined with CA, the treatment limited decay incidence in both naturally and artificially infected grapes	[27]
		Fumigation	50–70%	Scarlotta	Combined with MAP (2% O_2 + 5% CO_2), the treatment was efficient in decay control but caused sensorial quality losses (intense stem browning, off-flavours perception)	[31]
	Ozone (O_3)	Fumigation	20 µL/L	Scarlotta	Combined with MAP (2% O_2 + 5% CO_2), the treatment controlled the concentration of acetaldehyde, preserved rachis chlorophyll content and skin colour; the cumulative decay incidence was also reduced	[31]
		Periodic fumigation	2 µL/L	Superior Seedless, Cardinal CL80, and Regina Victoria	The treatment increased resveratrol content but led to low scores in sensory evaluation; high weight loss was also reported	[60]

Other chemicals are widely used as dipping solutions to sanitize fruit surfaces. The treatment of grapes by immersion or spraying with solutions of different generally recognized as safe (GRAS) salts at 1% reduced the percentage of spoiled fruit. This was the case with iron sulphate ($FeSO_4$) (92%), ammonium bicarbonate (NH_4HCO_3) (91%), sodium silicate (Na_2SiO_3) (89%), sodium bicarbonate ($NaHCO_3$) (76%) and sodium carbonate (Na_2CO_3) (74%) (application in pre-harvest, decay measured post-harvest) [55]. However, treatment with $FeSO_4$ could cause small black spots on the grape surface [55]. Disinfection by dipping in 32% ethanol, followed by six weeks of cold storage, reduced natural decay incidence on 'Scarlotta Seedless' from about 60% to 4.1% [54]. Nevertheless, the use of large quantities of ethanol is expensive and may be dangerous, due to its flammability. A more practical method is the use of ethanol vapour-generating bags, that confer longer protection, effectively reducing decay incidence in artificially inoculated grapes stored for one month, in a comparable way to SO_2 generating-pads in polyethylene bags [56]. In this case, significantly lower weight loss and moderate stem browning were also observed [56]. Furthermore, it is relevant to underline that active coatings associated with selected films represent a promising strategy to increase table grape shelf life [69].

Recently, Gorrasi et al. [70] demonstrated the efficacy of active packaging based on a food grade acrylic resin filled with Layered Double Hydroxide (LDH) nanofiller hosting antimicrobial 2-acetoxybenzoic anion (salicylate), on microbial control during table grape (cv Egnathia) storage.

In addition to ethanol vapours, other gas types have been used as fumigation treatment for the sanitization of bunches. With this scope, chlorine dioxide (ClO_2) is a gaseous disinfectant admitted in the sanitization of uncut and unpeeled fruits and vegetables. In a recent study, Chen et al. [57] reported a reduction of decay incidence and of rachis browning in table grapes treated with ClO_2 during storage. The Food and Drug Administration (FDA) has approved ClO_2, given that these treatments might leave chlorite residues on food products at non-hazardous concentrations [71]. Nitrous oxide (N_2O) is another gas tested to control post-harvest decay in fruit crops. In vitro tests did not show inhibition against grey mould; however, in vivo experiments in table grapes fumigated for 6 h showed a significant reduction in decay development during six days of cold storage [58]. Therefore, it was hypothesized that N_2O was indirectly able to inhibit grey mould by increasing the host's disease resistance [58].

The use of pre-treatments with high concentrations of CO_2 have been widely studied; these showed great potential in decay control and prevention of water loss and oxidative damage [59]. In Cardinal table grapes, these effects seem to be related to the specific induction of defence proteins, including dehydrins and proteins associated with pathogenesis, as well as endogenous protective osmolytes [59]. In the last few years, different concentrations of CO_2 were evaluated. Pre-treatments with 20% of CO_2 for three days [59], 40% CO_2 for 48 h followed by CA storage [27], and 50–70% for 24 h followed by MAP [31], were all effective against post-harvest decay of the cultivars assayed. Although all the treatments guaranteed basic quality standards for commercial table grapes, a concentration-dependent effect has been observed. However, as previously mentioned, the use of pre-storage application of a high concentration of CO_2 causes cultivar-dependent collateral effects such as rachis, berries browning and off-flavours [43].

Ozone fumigation is one of the most prominent sanitation strategies for fruits and vegetables [72,73]. Different approaches have been developed for ozone-based treatments on table grapes [74,75]. Among these, continuous exposure in controlled atmosphere during cold storage has been reported [28,29]. Decay reduction was confirmed only with pre-treatment at 20 µL/L for 30 min, followed by MAP storage [31]. Interestingly, intermittent ozone treatment (2 µL/L, 12 h for day) induced higher resveratrol accumulation (in three different table grape cultivars) [60]. Moreover, this could be responsible for decreases in the level of pesticide residues (phenomena reported for grapes stored in ozone atmosphere) [75,76]. Nevertheless, ozone is corrosive and represents a worker hazard [77], and, among the quality parameters, significant weight loss during storage was usually highlighted [28,44,60].

4. Biological Methods to Control *B. cinerea* in Table Grapes

Consumers widely accept the development of bio-based applications to exert microbial control in agro-food chains because of the growing demand for eco-friendly approaches and products free of synthetic chemicals [78–80]. For these purposes, several protective cultures [81–84] and compounds of biological origin [80,85] have been assessed for their possible use as Biological Control Agents (BCAs) against *B. cinerea* in table grapes.

4.1. Microbial Resources

Several yeast species are found in association with the surface of the grapes, in particular, the genera *Saccharomyces, Candida, Dekkera, Pichia, Hanseniaspora, Metschnikowia, Kluyveromyces, Saccharomycodes, Schizosaccharomyces, Torulaspora,* and *Zygosaccharomyces* [86,87]. Highly variable in terms of relative proportion, often as a function of the sanitary condition of the grapes, these species have different significances in oenology, i.e., pro-technological, spoilage, biocontrol, production of toxic catabolites [88–92]. On the other hand, it is possible to find prokaryotic organisms present on the grape surface that exert their biotechnological action in the last phases of the winemaking process [93]. This broad microbial diversity justifies massive isolation of yeasts and bacteria to preserve and characterize strains of biotechnological interest [94–96]. This isolation can be of microorganisms from plants, grape bunches, musts or wines and selection is made of those capable of inhibiting undesired microbe development on grapevines [97,98] up to the final steps of wine production [99]. This reservoir of microbial-based biocontrol solutions has also been exploited in fruits [100–103], in several cases offering the option to inhibit *B. cinerea* in table grapes (Table 3).

Among yeast species, strains belonging to *Saccharomyces* are the most commonly studied because of their pivotal function in alcoholic fermentation and their role as a biological model organism [117–119]. Recently, Nally et al. [108] used a fruit decay test on wounded table grape berries to screen the activity of 65 yeasts, previously tested against *B. cinerea* by using in vitro approaches. They found that 15 *S. cerevisiae* strains and one strain of *Sch. pombe*, isolated from grape must, were able to reduce grey mould decay [108]. Among these, the disease incidence of grapes treated with *Sch. pombe* BSchp67 reached 29.9%, while 9 strains of *S. cerevisiae* were able to fully inhibit decay development when added at a concentration of 10^7 cells/mL [108].

Regarding the non-*Saccharomyces* yeasts, *H. uvarum* is a species of enological interest, usually present on the grape surface [120,121]. In various studies, it has demonstrated an antagonistic property, mainly based on competition for living space [122]. The addition of this yeast has been implicated in the reduced incidence of grey mould disease in artificially inoculated table grapes [111]. Moreover, this antagonistic activity was enhanced by the addition in the formulation of salicylic acid or salts, such as sodium bicarbonate or ammonium molybdate [109,123]. *Starmerella bacillaris* (synonym *Candida zemplinina*) is another species of interest, commonly isolated from grapevines/musts [124,125] and from wines fermented by using botrytized grapes [126,127]. Three *Starm. bacillaris* strains, recently isolated from these wines, denoted a significative antifungal activity, probably addressable to the release of volatile organic compounds (VOCs) [110]. The production of VOCs is widely diffused among yeasts. Mewa-Ngongang et al. [112] observed a fungicidal effect of *C. pyralidae* Y1117 and *P. kluyveri* Y1125, mediated by VOC release in a closed environment, able to inhibit fungal growth for five weeks of storage [112].

Table 3. Main microbial strains investigated in the last ten years against grey mould decay in table grapes. Where possible, Inhibition Percentage (IP), Disease Incidence (DI), and Disease Reduction (DR) were reported to quantify the activity of each strain.

	Microbial Strain	Source of Isolation	Activity	Cultivar Tested	Ref.
Yeasts	*Issatchenkia terricola* 156a5	Thompson seedless	IP = ~80%	Flame seedless	[104]
	Wickerhamomyces anomalus BS91		DI = ~50%		
	Metschnikowia pulcherrima MPR3	Fermented olive and pomegranate	DI = 6.7%	Not specified	[105,106]
	Aureobasidium pullulans PI1		DI = ~55%		
	Meyerozyma guilliermondii Ka21, Kh59	Thompson seedless	IP = 47.6%	Thompson seedless	[107]
	Candida membranifaciens Kh69		IP = ~42%		
	Saccharomyces cerevisiae spp. (9 strains)	Grape must	DI = 0%	Red globe	[108]
	Schizosaccharomyces pombe BSchp67		DI = 29.92%		
	Hanseniaspora uvarum SEHMA61	Wild grape	-	Not specified	[109]
	Pichia kluyveri SEHMA6B		-		
	Starmerella bacillaris PAS151	Ripe grape must	DR = ~40%	Not specified	[110]
	Hanseniaspora uvarum	Strawberry	DI = 51,8%	Kyoho	[111]
	Candida pyralidae Y1117	Grape must	DI = 0%	Regal seedless	[112,113]
	Pichia kluyveri Y1125	*Sclerocarya birrea* juice	DI = 0%		
Bacteria	*Bacillus* sp. Kh26	Thompson seedless	IP = 49.9%	Thompson seedless	[107]
	Ralstonia sp. N1		IP = 54.7%		
	Bacillus amyloliquefaciens NCPSJ7	Ginger field	DI = 36%	Red globe	[114]
	Bacillus amyloliquefaciens RS-25	Jujube fruit	DR = 86.6%	Red globe	[115]
	Bacillus licheniformis MG-4	Strawberry	DR = 84.7%		
	Bacillus subtilis Pnf-4	Wheat plant	DR = 69.95%		
	Bacillus subtilis Z-14	Wheat soil	DR = 42.43%		
	Paenibacillus pasadenensis R16	Barbera	DR = 27.5%	Black magic	[116]

In vivo studies demonstrated that grey mould can be efficiently controlled by various microbial antagonists isolated from a large variety of vegetal matrices. *Wickerhamomyces anomalus* BS91, *M. pulcherrima* MPR3, and *Aureobasidium pullulans* PI1 were isolated from spontaneous olive fermentation and pomegranate, minimally processed. In detail, *M. pulcherrima* strain showed the best antifungal activity (disease incidence (DI) = 6.7%, disease severity (DS) = 2.7%), followed by *W. anomalus* BS91 and *A. pullulans* PI1, and all of these yeasts were capable of VOC production [106]. In particular, the antagonistic activity of *W. anomalus* seemed to be connected to a killer phenotype [106]. Enzyme secretion in the environment, such as b-1,3-glucanase, pectinase, and protease, was also reported for *W. anomalus* and *A. pullulans* [106], whereas, the activity of *M. pulcherrima* was probably associated with iron depletion [128]. In the patenting literature, two patents based on *M. fructicola* strain's biocontrol applications for viticultural applications have been reported [129].

Epiphytic *Issatchenkia terricola* yeasts isolated from 'Thompson Seedless' grapes' surface have shown the ability to reduce decay caused by *B. cinerea* up to 80% compared to the untreated control [104]. In another study, yeast and bacteria strains were isolated from fruits and leaves of the same cultivar without any signs of infection, and tested for potential applications in biocontrol [107]. Yeasts were identified as *Candida membranifasciens* Kh69 and *Meyerozyma guilliermondii* Ka21 and Kh59, while bacteria were *Bacillus* spp. Kh26 and *Ralstonia* spp. N1. All tested microbes were able to increase *B. cinerea* inhibition from 23.8% to 54.7%. Among these, the highest level was found for *Ralstonia* spp. N1(54.7%), while *Bacillus* spp. Kh26 and *M. guilliermondii* Ka21 and Kh59 showed inhibition below 50% [107].

Still on the prokaryotic side, a bacterial strain, *Paenibacillus pasadenensis* R16, isolated from grapevine cultivar 'Barbera', has shown a reduction in disease incidence of grey mould by 27.5% [116]. It was also supposed that the main metabolite responsible for antifungal activity was farnesol which was never before reported to have biocontrol potential [116]. A large number of bacterial strains belonging to *Bacillus* spp. are reported to have antimicrobial activity against several plant phytopathogens [130–132]. In fact, a lot of commercial bio-fungicides, such as *B. subtilis* QST713 (Serenade®, Bayer CropScience) and *B. amyloliquefaciens* FZB24 (Taegro®, Novozymes), are now available and effective against grey mould on grapes. Recently, Chen et al. [115] demonstrated the ability of four *Bacillus* strains, isolated from various ecological niches, to control decay development in table grapes and other fruit crops. The most vigorous antifungal activity was recorded in *B. subtilis* Z-14 [115]. VOC production, enzyme, siderophores, and lipopeptide antibiotics were proposed as possible modes of action.

4.2. Antimicrobial Compounds of Biological Origin

Recently, there have been intense investigations conducted in the field of natural antimicrobials and their effectiveness. Many biological compounds have been tested for the bio-control of table grape spoilages. These compounds include classes of chemicals/matrices such as vegetal extracts, essential oils, and defence inducers (Table 4).

Among the vegetal compounds, volatiles generated from cellulose soaked with garlic hydro-alcoholic extract and its derived sulfur compounds have shown anti-grey mould activity in packaged table grapes both at 4 and 25 °C, during the 14 days of experimental trials [133]. However, organoleptic and sensorial adverse effects of this treatment have still not been investigated [133]. Cinnamic acid, extracted from cinnamon bark, is widely used as a food additive. Dipping the berries in a solution of 10 mM cinnamic acid can significantly decrease the incidence of decay development up to half of that in control after four days of storage at 25 °C [134]. Hinokitiol is a natural monoterpenoid mainly extracted from the wood of *Cupressaceae*. In a recent study [135], no decay was visible after 60 h at 22 °C in artificially wounded/inoculated table grape berries treated with a 3 g/L hinokitiol solution [135].

Table 4. Main biological compounds investigated in the last ten years against grey mould decay on table grapes.

	Biological Compounds	Concentration	Treatment	Cultivar	Effects	Ref.
Vegetal extract	Hydro-alcoholic garlic extract and derived sulfur compounds	2 mL and 20 µL	Volatiles release	Flame Seedless	The treatment efficiently controlled the decay in packed grapes at 4 and 25 °C for 14 days	[133]
	Cinnamic acid	10 mM	Dipping	Manai	The treatment halved the decay incidence after four days at 25 °C	[134]
	Hinokitiol	3 g/L	Wound inoculation	Manai	No visible decay was reported after 60 h at 22 °C	[135]
Essential Oil	Mint EO	500 µL/L	Volatiles release	Not specified	Reduction of decay in packed grapes	[136]
Other compounds	Methyl jasmonate	10 µmol/L	Volatiles release	Kyoho	Reduction of the decay incidence	[137]
	Fulvic acid	20 mg/mL	Dipping	Mare's milk	Induction of resistance mainly through the activation of phenylpropanoid pathway	[138]
	Pterostilbene and Piceatannol	50 mg/L	Wound inoculation	Mare's milk	Reduction of disease incidence and severity	[139]
	Putrescine	1–2 mM	Dipping	Michele Palieri	Combined with ultrasound, the treatment maintained high levels of anthocyanins, total phenolic content, antioxidant capacity, sensory acceptability and reduced decay incidence during storage	[23]
Edible coating	Chitosan	-	Coating	Crimson	Combined with UV-C irradiation, the treatment increased the resveratrol content, maintained sensorial quality, and reduced fungal decay	[24]
	Chitosan/Silica polymer	0.5–1%	Spraying	Italia	The treatment reduced natural infection; no adverse effect in terms of quality (titratable acidity [TA], total soluble solids [TSS], berry color, mass loss, stem browning and shattered berries) was observed	[140]
	Chitosan + *Salvia fruticosa* Extract	500 mg/L (SE)	Dipping	Thompson Seedless	Control efficacy comparable to thiabendazole, decreased the weight loss during cold storage, preserved TSS and TA	[141]

Table 4. *Cont.*

Biological Compounds	Concentration	Treatment	Cultivar	Effects	Ref.
Chitosan + Mint Essential Oil	1.25–5 µL/mL (MEO)	Dipping	Isabella	The treatment delayed the decay development and reduced incidence; color and firmness were enhanced, did not negatively affect TSS and TA	[142]
Alginate + Vanillin	0.5–1.5% (V)	Spraying	Lavalleé and Razaki	Reduction of natural yeasts and mould growth, prevention of weight and firmness losses. TSS, TA, and color showed minor changes compared to control grapes.	[143]

Essential oils (EOs) from many plants, such as thymus and lemongrass, have revealed great potential in post-harvest disease control [144]. In addition, the effect of mint EOs was recently investigated by using direct contact (e.g., dipping) and volatile methods (filter paper) [136]. In this study, EO released by the paper was more effective than the direct contact and was capable of inhibiting *B. cinerea* in artificially inoculated trials during nine days of shelf-life [136]. However, the effect on product flavour and consumer acceptance was not investigated.

Another research field involves the use of vegetal hormones, plant activators, and inner signalling molecules. These molecules act through a complex signalling network under the control of salicylic acid, ethylene, jasmonic acid, and phenylpropanoid pathways, which leads to the increase of specific secondary metabolites (e.g., flavonoids, soluble sugars, and phytoalexins). Methyl jasmonate is a volatile compound that mediates stress responses in plants and has shown to promote fungal resistance in various fruit crops. Recently, it was found to be effective in lessening the development of *B. cinerea* in artificially infected table grapes [137]. In this study, the fruits were packed in the presence of a filter paper soaked with a solution of methyl jasmonate at 10 μmol/L and stored at 25 °C [137]. The disease incidence in the treated fruits after 24, 36, and 48 h was 41.7%, 60.6%, and 86.5% of that in the control trial, respectively [137].

Fulvic acids (FA) are the soluble fraction of natural organic matter and are used in agriculture as a plant growth promoter and to control several plant diseases. Xu et al. [138] assayed different concentrations of FA as dipping solutions for wounded table grape fruits, subsequently sprinkled with a conidia suspension of *B. cinerea*. After six days of incubation at 22 °C, the treatment with a solution at 20 mg/mL FA was found to be effective by reducing decay development [138]. The authors suggested that secondary metabolites produced by the berry mediate antifungal activity. However, the formation of necrotic spots was reported [138].

Among secondary metabolites, phytoalexins are synthesized by the plants as broad-spectrum inhibitors. Stilbenoids, including pterostilbene and piceatannol, are phytoalexins commonly found in vine leaves and wine [139]. "Mare's milk" table grapes treated with 50 mg/L pterostilbene did not show any sign of infection while piceatannol at the same concentration reduced grey mould disease by 75% after nine days storage at 22 °C [139]. These molecules seemed to be the most effective in a group of seven phenolic compounds, including resveratrol and coumarin [139].

Edible coatings made with natural polymers like chitosan or alginate can act as a cover material able to wrap the berry. Thus, these formulations can extend the shelf-life of fruit crops and maintain quality reducing water losses [145,146]. Chitosan is a linear polysaccharide composed of D-glucosamine and N-acetyl-D-glucosamine linked by a β-(1→4) bond obtained by treating the exoskeleton of arthropods with alkaline solutions. Recently, it was found that chitosan-silica nanocomposite polymers can reduce the incidence of decay in grape berries by 59% [140]. Moreover, this coating can be used to incorporate bioactive compounds. An additive effect of chitosan combined with *Salvia fruticosa* Mill. extract [141] and *Mentha piperita* or *M. villosa* essential oil [142] was reported. Alginate is another biocompatible and biodegradable polymer extracted from brown algae and used as a food additive with the code E401. It was demonstrated that the incorporation of vanillin, a phenolic compound, in a coating formulation prolongs the shelf life of table grapes until 35 days of storage, by reducing total yeasts and mould counts [143]. However, the retention of soluble solids, titratable acidity, firmness, and color was also enhanced.

5. Conclusions and Future Directions

Post-harvest fungal decay of fruits and vegetables is responsible for huge levels of economic loss and account consistently for large quantities of agro-food waste [147–150]. To improve economic, social, and environmental sustainability in the sector of table grapes, this review paper provides an overview of the wide plethora of physical, chemical, and bio-based solutions to improve the control of fungal pathogens and spoilage fungi. Each treatment has peculiar benefits and limitations that affect the concrete applications and shape different future perspectives [151]. For example, considering limitations, ozone does not always penetrate natural openings efficiently; condensation inside the package (MAP) increases the chance of microbial decay of produce; the antagonistic target of a biocontrol agent can have a strain-dependent spectrum. In some cases, the limitation is due to lack of harmonization of regulations and consumer acceptance (e.g., irradiation), and investment needs compared to the volume of production (e.g., CA storage) rather than of specific technological or biological issues [151].

As in other fields of food technology, an integrated management program (combining two or more different solutions) could be useful to minimize post-harvest losses caused by undesired fungal development [147,152–155]. Synergistic approaches have also been developed to reduce *B. cinerea* incidence in table grapes, adopting hurdles technology [23,24,27,31,51]. In other cases, one treatment aimed to reduce microbial contamination, while another was applied to stabilize fruit quality and the microbial population during cold storage and/or shelf-life [27,31,156]. Moreover, it is important to underline that a consistent range of solutions has been developed and tested on other fruits and vegetable [157–163] and, in several cases, could be tested/transferred for application on table grapes. Among the other green solutions, poorly explored in grapes, is the exploitation of lactic acid bacteria as biocontrol agents [164,165]: prokaryotic organisms that received interest also in the light of additional positive side effects, e.g., probiotic activity and antagonistic activity against food-borne pathogens [166–170].

Author Contributions: Investigation, N.D.S., B.P., F.G., V.C., G.C., G.S. and P.R.; Conceptualization, N.D.S., B.P., F.G., M.C., V.T., V.S., V.C., G.C., G.S. and P.R.; Literature Search, N.D.S., B.P., F.G., M.C., V.T., V.S., V.C., G.C., G.S. and P.R.; Writing—Original Draft Preparation, N.D.S., V.C. and P.R.; Writing—Review and Editing, B.P., F.G., M.C., V.T., V.S., G.C. and G.S. All authors have read and agreed to the published version of the manuscript.

Funding: ABA MEDITERRANEA SCA was funded through Piani Operativi 2020 REG UE N 1308/13, REG UE N 2017/891, REG UE N 2017/892.

Acknowledgments: Pasquale Russo is the beneficiary of a grant by MIUR in the framework of 'AIM: Attraction and International Mobility' (PON R&I2014-2020) (practice code D74I18000190001). The authors acknowledge (i) the two anonymous reviewers for their suggestions and comments, (ii) Massimo Franchi and Francesco De Marzo of the Institute of Sciences of Food Production—CNR for the skilled technical support provided during the realization of this work and (iii) Sergio Pelosi of the Institute of Sciences of Food Production—CNR for the critical reading.

Conflicts of Interest: The authors declare no conflict of interest.

References

1. Pezzuto, J.M. Grapes and human health: A perspective. *J. Agric. Food Chem.* **2008**, *56*, 6777–6784. [CrossRef] [PubMed]

2. FAO OIV. Table and Dried Grapes: World Data Available. Available online: http://www.fao.org/documents/card/en/c/709ef071-6082-4434-91bf-4bc5b01380c6/ (accessed on 12 August 2020).

3. Ahmed, S.; Roberto, S.; Domingues, A.; Shahab, M.; Junior, O.; Sumida, C.; de Souza, R. Effects of Different Sulfur Dioxide Pads on Botrytis Mold in 'Italia' Table Grapes under Cold Storage. *Horticulturae* **2018**, *4*, 29. [CrossRef]

4. Williamson, B.; Tudzynski, B.; Tudzynski, P.; Van Kan, J.A.L. Botrytis cinerea: The cause of grey mould disease. *Mol. Plant Pathol.* **2007**, *8*, 561–580. [CrossRef] [PubMed]

5. Dean, R.; Van Kan, J.A.; Pretorius, Z.A.; Hammond-Kosack, K.E.; Di Pietro, A.; Spanu, P.D.; Rudd, J.J.; Dickman, M.; Kahmann, R.; Ellis, J.; et al. The Top 10 fungal pathogens in molecular plant pathology. *Mol. Plant Pathol.* **2012**, *13*, 414–430. [CrossRef] [PubMed]

6. Droby, S.; Lichter, A. Post-harvest Botrytis infection: Etiology, development and management. In *Botrytis: Biology, Pathology and Control*; Springer: Berlin/Heidelberg, Germany, 2007; pp. 349–367.

7. Domingues, A.; Roberto, S.; Ahmed, S.; Shahab, M.; José Chaves Junior, O.; Sumida, C.; de Souza, R. Postharvest Techniques to Prevent the Incidence of *Botrytis* Mold of 'BRS Vitoria' Seedless Grape under Cold Storage. *Horticulturae* **2018**, *4*, 17. [CrossRef]

8. Capozzi, V.; Fiocco, D.; Amodio, M.L.; Gallone, A.; Spano, G. Bacterial Stressors in Minimally Processed Food. *Int. J. Mol. Sci.* **2009**, *10*, 3076–3105. [CrossRef] [PubMed]

9. Francis, G.A.; Gallone, A.; Nychas, G.J.; Sofos, J.N.; Colelli, G.; Amodio, M.L.; Spano, G. Factors Affecting Quality and Safety of Fresh-Cut Produce. *Crit. Rev. Food Sci. Nutr.* **2012**, *52*, 595–610. [CrossRef]

10. Cavallo, D.P.; Cefola, M.; Pace, B.; Logrieco, A.F.; Attolico, G. Non-destructive and contactless quality evaluation of table grapes by a computer vision system. *Comput. Electron. Agric.* **2019**, *156*, 558–564. [CrossRef]

11. Ferrara, G.; Gallotta, A.; Pacucci, C.; Matarrese, A.M.S.; Mazzeo, A.; Giancaspro, A.; Gadaleta, A.; Piazzolla, F.; Colelli, G. The table grape 'Victoria' with a long shaped berry: A potential mutation with attractive characteristics for consumers. *J. Sci. Food Agric.* **2017**, *97*, 5398–5405. [CrossRef]

12. Piazzolla, F.; Pati, S.; Amodio, M.L.; Colelli, G. Effect of harvest time on table grape quality during on-vine storage. *J. Sci. Food Agric.* **2016**, *96*, 131–139. [CrossRef]

13. Iglesias-Carres, L.; Mas-Capdevila, A.; Bravo, F.I.; Aragonès, G.; Arola-Arnal, A.; Muguerza, B. A comparative study on the bioavailability of phenolic compounds from organic and nonorganic red grapes. *Food Chem.* **2019**, *299*, 125092. [CrossRef] [PubMed]

14. Council regulation (EC) No 834/2007 of 28 June 2007 on organic production and labelling of organic products and repealing regulation (EEC) No 2092/1. *Off. J. Eur. Union* **2007**, 1–23.

15. Willer, H.; Schaack, D.; Lernoud, J. Organic Farming and Market Development in Europe and the European Union. In *The World of Organic Agriculture—Statistics and Emerging Trends 2019*; Willer, H., Lernoud, J., Eds.; Research Institute of Organic Agriculture FiBL and IFOAM—Organics International, Frick and Bonn: Rheinbreitbach, Germany, 2019; pp. 217–254.

16. Romanazzi, G.; Lichter, A.; Gabler, F.M.; Smilanick, J.L. Recent advances on the use of natural and safe alternatives to conventional methods to control postharvest gray mold of table grapes. *Postharvest Biol. Technol.* **2012**, *63*, 141–147. [CrossRef]

17. Lichter, A.; Kaplunov, T.; Zutahy, Y.; Lurie, S. Unique techniques developed in Israel for short- and long-term storage of table grapes. *Isr. J. Plant Sci.* **2016**, *63*, 2–6. [CrossRef]

18. Sonker, N.; Pandey, A.K.; Singh, P. Strategies to control post-harvest diseases of table grape: A review. *J. Wine Res.* **2016**, *27*, 105–122. [CrossRef]

19. Youssef, K.; Roberto, S.R.; Chiarotti, F.; Koyama, R.; Hussain, I.; de Souza, R.T. Control of *Botrytis* mold of the new seedless grape 'BRS Vitoria' during cold storage. *Sci. Hortic.* **2015**, *193*, 316–321. [CrossRef]

20. Crisosto, C.H.; Mitchell, F.G. Postharvest Handling Systems: Table grapes. In *Postharvest Technology of Horticultural Crops*; Kader, A.A., Ed.; University of California Agricultural and Natural Resources Pub: Davis, CA, USA, 2002; pp. 357–363.

21. Sabir, F.K.; Sabir, A. Quality response of table grapes (*Vitis vinifera* L.) during cold storage to postharvest cap stem excision and hot water treatments. *Int. J. Food Sci. Technol.* **2013**, *48*, 999–1006. [CrossRef]

22. Chiabrando, V.; Giacalone, G. Efficacy of hot water treatment as sanitizer for minimally processed table grape. *J. Clean. Prod.* **2020**, *257*, 120364. [CrossRef]

23. Bal, E.; Kok, D.; Torcuk, A.I. Postharvest putrescine and ultrasound treatments to improve quality and postharvest life of table grapes (*Vitis vinifera* L.) cv. Michele Palieri. *J. Cent. Eur. Agric.* **2017**, *18*. [CrossRef]

24. Freitas, P.M.; López-Gálvez, F.; Tudela, J.A.; Gil, M.I.; Allende, A. Postharvest treatment of table grapes with ultraviolet-C and chitosan coating preserves quality and increases stilbene content. *Postharvest Biol. Technol.* **2015**, *105*, 51–57. [CrossRef]

25. Romanazzi, G.; Nigro, F.; Ippolito, A. Effectiveness of a short hyperbaric treatment to control postharvest decay of sweet cherries and table grapes. *Postharvest Biol. Technol.* **2008**, *49*, 440–442. [CrossRef]

26. Guentzel, J.L.; Lam, K.L.; Callan, M.A.; Emmons, S.A.; Dunham, V.L. Postharvest management of gray mold and brown rot on surfaces of peaches and grapes using electrolyzed oxidizing water. *Int. J. Food Microbiol.* **2010**, *143*, 54–60. [CrossRef] [PubMed]

27. Teles, C.S.; Benedetti, B.C.; Gubler, W.D.; Crisosto, C.H. Prestorage application of high carbon dioxide combined with controlled atmosphere storage as a dual approach to control *Botrytis cinerea* in organic 'Flame Seedless' and 'Crimson Seedless' table grapes. *Postharvest Biol. Technol.* **2014**, *89*, 32–39. [CrossRef]

28. Vlassi, E.; Vlachos, P.; Kornaros, M. Effect of ozonation on table grapes preservation in cold storage. *J. Food Sci. Technol.* **2018**, *55*, 2031–2038. [CrossRef] [PubMed]

29. Feliziani, E.; Romanazzi, G.; Smilanick, J.L. Application of low concentrations of ozone during the cold storage of table grapes. *Postharvest Biol. Technol.* **2014**, *93*, 38–48. [CrossRef]

30. Liguori, G.; Sortino, G.; De Pasquale, C.; Inglese, P. Effects of modified atmosphere packaging on quality parameters of minimally processed table grapes during cold storage. *Adv. Hortic. Sci.* **2015**, *29*, 152–154.

31. Admane, N.; Genovese, F.; Altieri, G.; Tauriello, A.; Trani, A.; Gambacorta, G.; Verrastro, V.; Di Renzo, G.C. Effect of ozone or carbon dioxide pre-treatment during long-term storage of organic table grapes with modified atmosphere packaging. *LWT* **2018**, *98*, 170–178. [CrossRef]

32. Cefola, M.; Pace, B. High CO2-modified atmosphere to preserve sensory and nutritional quality of organic table grape (cv. 'Italia') during storage and shelf-life. *Eur. J. Hortic. Sci.* **2016**, *81*, 197–203. [CrossRef]

33. Vilaplana, R.; Chicaiza, G.; Vaca, C.; Valencia-Chamorro, S. Combination of hot water treatment and chitosan coating to control anthracnose in papaya (*Carica papaya* L.) during the postharvest period. *Crop Prot.* **2020**, *128*, 105007. [CrossRef]

34. Kahramanoğlu, İ.; Chen, C.; Chen, Y.; Chen, J.; Gan, Z.; Wan, C. Improving Storability of "nanfeng" Mandarins by Treating with Postharvest Hot Water Dipping. *J. Food Qual.* **2020**, *2020*. [CrossRef]

35. Nigro, F.; Ippolito, A.; Lima, G. Use of UV-C light to reduce *Botrytis* storage rot of table grapes. *Postharvest Biol. Technol.* **1998**, *13*, 171–181. [CrossRef]

36. Amodio, M.L.; Rinaldi, R.; Colelli, G. Influence of atmosphere composition on quality attributes of ready-to-cook fresh-cut vegetable soup. In Proceedings of the IV International Conference on Managing Quality in Chains-The Integrated View on Fruits and Vegetables Quality, Bangkok, Thailand, 7–10 August 2006; pp. 677–684.

37. Daş, E.; Gürakan, G.C.; Bayındırlı, A. Effect of controlled atmosphere storage, modified atmosphere packaging and gaseous ozone treatment on the survival of *Salmonella* Enteritidis on cherry tomatoes. *Food Microbiol.* **2006**, *23*, 430–438. [CrossRef]

38. Cefola, M.; Pace, B.; Buttaro, D.; Santamaria, P.; Serio, F. Postharvest evaluation of soilless-grown table grape during storage in modified atmosphere. *J. Sci. Food Agric.* **2011**, *91*, 2153–2159. [CrossRef] [PubMed]

39. Cefola, M.; Renna, M.; Pace, B. Marketability of ready-to-eat cactus pear as affected by temperature and modified atmosphere. *J. Food Sci. Technol.* **2014**, *51*, 25–33. [CrossRef]

40. La Zazzera, M.; Amodio, M.L.; Colelli, G. Designing a modified atmosphere packaging (MAP) for fresh-cut artichokes. *Adv. Hortic. Sci.* **2015**, *29*, 24–29.

41. Retamales, J.; Defilippi, B.G.; Arias, M.; Castillo, P.; Manríquez, D. High-CO2 controlled atmospheres reduce decay incidence in Thompson Seedless and Red Globe table grapes. *Postharvest Biol. Technol.* **2003**, *29*, 177–182. [CrossRef]

42. Artés-Hernández, F.; Aguayo, E.; Artés, F. Alternative atmosphere treatments for keeping quality of 'Autumn seedless' table grapes during long-term cold storage. *Postharvest Biol. Technol.* **2004**, *31*, 59–67. [CrossRef]

43. Crisosto, C.H.; Garner, D.; Crisosto, G. High Carbon Dioxide Atmospheres Affect Stored "Thompson Seedless" Table Grapes. *HortScience* **2002**, *37*, 1074–1078. [CrossRef]

44. Palou, L.; Crisosto, C.H.; Smilanick, J.L.; Adaskaveg, J.E.; Zoffoli, J.P. Effects of continuous 0.3 ppm ozone exposure on decay development and physiological responses of peaches and table grapes in cold storage. *Postharvest Biol. Technol.* **2002**, *24*, 39–48. [CrossRef]

45. Carter, M.Q.; Chapman, M.H.; Gabler, F.; Brandl, M.T. Effect of sulfur dioxide fumigation on survival of foodborne pathogens on table grapes under standard storage temperature. *Food Microbiol.* **2015**, *49*, 189–196. [CrossRef]

46. Gao, H.; Hu, X.; Zhang, H.; Wang, S.; Liu, L. Study on sensitivity of table grapes to SO$_2$. In Proceedings of the XXVI International Horticultural Congress: Issues and Advances in Postharvest Horticulture, Toronto, ON, Canada, 11–17 August 2002; pp. 541–548.

47. Zoffoli, J.P.; Latorre, B.A.; Naranjo, P. Hairline, a postharvest cracking disorder in table grapes induced by sulfur dioxide. *Postharvest Biol. Technol.* **2008**, *47*, 90–97. [CrossRef]

48. Lou, T.; Huang, W.; Wu, X.; Wang, M.; Zhou, L.; Lu, B.; Zheng, L.; Hu, Y. Monitoring, exposure and risk assessment of sulfur dioxide residues in fresh or dried fruits and vegetables in China. *Food Addit. Contam. Part A* **2017**, *34*, 918–927. [CrossRef] [PubMed]

49. EPA. Pesticide tolerance for sulfur dioxide. *The Federal Register*, 10 May 1989; pp. 384–385.

50. European Commission. Regulation (EC) No 1333/2008 of the European Parliament and of the Council of 16 December 2008 on food additives. *Off. J. Eur. Community* **2008**, *50*, 18.

51. Xu, D.; Yu, G.; Xi, P.; Kong, X.; Wang, Q.; Gao, L.; Jiang, Z. Synergistic Effects of Resveratrol and Pyrimethanil against Botrytis cinerea on Grape. *Molecules* **2018**, *23*, 1455. [CrossRef] [PubMed]

52. Vitale, A.; Panebianco, A.; Polizzi, G. Baseline sensitivity and efficacy of fluopyram against *Botrytis cinerea* from table grape in Italy. *Ann. Appl. Biol.* **2016**, *169*, 36–45. [CrossRef]

53. Youssef, K.; Roberto, S.; Colombo, R.; Canteri, M.; Elsalam, K. Acibenzolar-S-methyl against Botrytis mold on table grapes in vitro and in vivo. *Agron. Sci. Biotechnol.* **2019**, *5*, 52. [CrossRef]

54. Lee, J.-S.; Kaplunov, T.; Zutahy, Y.; Daus, A.; Alkan, N.; Lichter, A. The significance of postharvest disinfection for prevention of internal decay of table grapes after storage. *Sci. Hortic.* **2015**, *192*, 346–349. [CrossRef]

55. Youssef, K.; Roberto, S.R. Salt strategies to control *Botrytis* mold of 'Benitaka' table grapes and to maintain fruit quality during storage. *Postharvest Biol. Technol.* **2014**, *95*, 95–102. [CrossRef]

56. Candir, E.; Ozdemir, A.E.; Kamiloglu, O.; Soylu, E.M.; Dilbaz, R.; Ustun, D. Modified atmosphere packaging and ethanol vapor to control decay of 'Red Globe' table grapes during storage. *Postharvest Biol. Technol.* **2012**, *63*, 98–106. [CrossRef]

57. Chen, S.; Wang, H.; Wang, R.; Fu, Q.; Zhang, W. Effect of gaseous chlorine dioxide (ClO2) with different concentrations and numbers of treatments on controlling berry decay and rachis browning of table grape. *J. Food Process. Preserv.* **2018**, *42*, e13662. [CrossRef]

58. Xu, J.; Zhang, Z.; Li, X.; Wei, J.; Wu, B. Effect of nitrous oxide against *Botrytis cinerea* and phenylpropanoid pathway metabolism in table grapes. *Sci. Hortic.* **2019**, *254*, 99–105. [CrossRef]

59. Vazquez-Hernandez, M.; Navarro, S.; Sanchez-Ballesta, M.T.; Merodio, C.; Escribano, M.I. Short-term high CO2 treatment reduces water loss and decay by modulating defense proteins and organic osmolytes in Cardinal table grape after cold storage and shelf-life. *Sci. Hortic.* **2018**, *234*, 27–35. [CrossRef]

60. Cayuela, J.A.; Vázquez, A.; Pérez, A.G.; García, J.M. Control of Table Grapes Postharvest Decay by Ozone Treatment and Resveratrol Induction: *Food Sci. Technol. Int.* **2010**. [CrossRef]

61. Muri, S.D.; van der Voet, H.; Boon, P.E.; van Klaveren, J.D.; Brüschweiler, B.J. Comparison of human health risks resulting from exposure to fungicides and mycotoxins via food. *Food Chem. Toxicol.* **2009**, *47*, 2963–2974. [CrossRef] [PubMed]

62. Meena, R.S.; Kumar, S.; Datta, R.; Lal, R.; Vijayakumar, V.; Brtnicky, M.; Sharma, M.P.; Yadav, G.S.; Jhariya, M.K.; Jangir, C.K.; et al. Impact of Agrochemicals on Soil Microbiota and Management: A Review. *Land* **2020**, *9*, 34. [CrossRef]

63. Russo, P.; Berbegal, C.; De Ceglie, C.; Grieco, F.; Spano, G.; Capozzi, V. Pesticide Residues and Stuck Fermentation in Wine: New Evidences Indicate the Urgent Need of Tailored Regulations. *Fermentation* **2019**, *5*, 23. [CrossRef]

64. El Ghaouth, A.; Wilson, C.; Wisniewski, M. Biologically-Based Alternatives to Synthetic Fungicides for the Control of Postharvest diseases of Fruit and Vegetables. In *Diseases of Fruits and Vegetables: Volume II: Diagnosis and Management*; Naqvi, S.A.M.H., Ed.; Springer: Dordrecht, The Netherlands, 2004; pp. 511–535, ISBN 978-1-4020-2607-2.

65. Latorre, B.A.; Spadaro, I.; Rioja, M.E. Occurrence of resistant strains of *Botrytis cinerea* to anilinopyrimidine fungicides in table grapes in Chile. *Crop Prot.* **2002**, *21*, 957–961. [CrossRef]

66. Avenot, H.F.; Michailides, T.J. Progress in understanding molecular mechanisms and evolution of resistance to succinate dehydrogenase inhibiting (SDHI) fungicides in phytopathogenic fungi. *Crop Prot.* **2010**, *29*, 643–651. [CrossRef]

67. Amiri, A.; Heath, S.M.; Peres, N.A. Resistance to fluopyram, fluxapyroxad, and penthiopyrad in *Botrytis cinerea* from strawberry. *Plant Dis.* **2014**, *98*, 532–539. [CrossRef]

68. Ge, Y.-H.; Bi, Y.; Li, Y.-C.; Wang, Y. Resistance of harvested fruits and vegetables to diseases induced by ASM and its mechanism. *Sci. Agric. Sin.* **2012**, *45*, 3357–3362.

69. Cefola, M.; Pace, B.; Bugatti, V.; Vittoria, V. Active coatings for food packaging: A new strategy for table grape storage. *Acta Hortic.* **2015**, *1071*, 121–127. [CrossRef]

70. Gorrasi, G.; Bugatti, V.; Vertuccio, L.; Vittoria, V.; Pace, B.; Cefola, M.; Quintieri, L.; Bernardo, P.; Clarizia, G. Active packaging for table grapes: Evaluation of antimicrobial performances of packaging for shelf life of the grapes under thermal stress. *Food Packag. Shelf Life* **2020**, *25*, 100545. [CrossRef]

71. Smith, D.J.; Ernst, W.; Herges, G.R. Chloroxyanion Residues in Cantaloupe and Tomatoes after Chlorine Dioxide Gas Sanitation. *J. Agric. Food Chem.* **2015**, *63*, 9640–9649. [CrossRef]

72. Miller, F.A.; Silva, C.L.; Brandão, T.R. A review on ozone-based treatments for fruit and vegetables preservation. *Food Eng. Rev.* **2013**, *5*, 77–106. [CrossRef]

73. Aslam, R.; Alam, M.S.; Saeed, P.A. Sanitization Potential of Ozone and Its Role in Postharvest Quality Management of Fruits and Vegetables. *Food Eng. Rev.* **2020**, *12*, 48–67. [CrossRef]

74. Ozkan, R.; Smilanick, J.L.; Karabulut, O.A. Toxicity of ozone gas to conidia of *Penicillium digitatum*, *Penicillium italicum*, and *Botrytis cinerea* and control of gray mold on table grapes. *Postharvest Biol. Technol.* **2011**, *60*, 47–51. [CrossRef]

75. Gabler, F.M.; Smilanick, J.L.; Mansour, M.F.; Karaca, H. Influence of fumigation with high concentrations of ozone gas on postharvest gray mold and fungicide residues on table grapes. *Postharvest Biol. Technol.* **2010**, *55*, 85–90. [CrossRef]

76. Karaca, H. The Effects of Ozone-Enriched Storage Atmosphere on Pesticide Residues and Physicochemical Properties of Table Grapes. *Ozone Sci. Eng.* **2019**, *41*, 404–414. [CrossRef]

77. Zhang, J.; Wei, Y.; Fang, Z. Ozone pollution: A major health hazard worldwide. *Front. Immunol.* **2019**, *10*, 2518. [CrossRef]

78. Russo, P.; Fares, C.; Longo, A.; Spano, G.; Capozzi, V. *Lactobacillus plantarum* with Broad Antifungal Activity as a Protective Starter Culture for Bread Production. *Foods* **2017**, *6*, 110. [CrossRef]

79. Linares-Morales, J.R.; Gutiérrez-Méndez, N.; Rivera-Chavira, B.E.; Pérez-Vega, S.B.; Nevárez-Moorillón, G.V. Biocontrol Processes in Fruits and Fresh Produce, the Use of Lactic Acid Bacteria as a Sustainable Option. *Front. Sustain. Food Syst.* **2018**, *2*, 50. [CrossRef]

80. Raveau, R.; Fontaine, J.; Lounès-Hadj Sahraoui, A. Essential Oils as Potential Alternative Biocontrol Products against Plant Pathogens and Weeds: A Review. *Foods* **2020**, *9*, 365. [CrossRef] [PubMed]

81. Ling, L.; Han, X.; Li, X.; Zhang, X.; Wang, H.; Zhang, L.; Cao, P.; Wu, Y.; Wang, X.; Zhao, J.; et al. A *Streptomyces* sp. NEAU-HV9: Isolation, Identification, and Potential as a Biocontrol Agent against *Ralstonia solanacearum* of Tomato Plants. *Microorganisms* **2020**, *8*, 351. [CrossRef] [PubMed]

82. Into, P.; Khunnamwong, P.; Jindamoragot, S.; Am-in, S.; Intanoo, W.; Limtong, S. Yeast Associated with Rice Phylloplane and Their Contribution to Control of Rice Sheath Blight Disease. *Microorganisms* **2020**, *8*, 362. [CrossRef] [PubMed]

83. Arena, M.P.; Russo, P.; Spano, G.; Capozzi, V. Exploration of the Microbial Biodiversity Associated with North Apulian Sourdoughs and the Effect of the Increasing Number of Inoculated Lactic Acid Bacteria Strains on the Biocontrol against Fungal Spoilage. *Fermentation* **2019**, *5*, 97. [CrossRef]

84. Arena, M.P.; Russo, P.; Spano, G.; Capozzi, V. From Microbial Ecology to Innovative Applications in Food Quality Improvements: The Case of Sourdough as a Model Matrix. *J—Multidiscip. Sci. J.* **2020**, *3*, 3. [CrossRef]

85. Gálvez, A.; Abriouel, H.; Benomar, N.; Lucas, R. Microbial antagonists to food-borne pathogens and biocontrol. *Curr. Opin. Biotechnol.* **2010**, *21*, 142–148. [CrossRef]

86. Berbegal, C.; Spano, G.; Tristezza, M.; Grieco, F.; Capozzi, V. Microbial Resources and Innovation in the Wine Production Sector. *S. Afr. J. Enol. Vitic.* **2017**, *38*, 156–166. [CrossRef]

87. Garofalo, C.; Russo, P.; Beneduce, L.; Massa, S.; Spano, G.; Capozzi, V. Non-*Saccharomyces* biodiversity in wine and the 'microbial terroir': A survey on Nero di Troia wine from the Apulian region, Italy. *Ann. Microbiol.* **2016**, *66*, 143–150. [CrossRef]

88. Benito, Á.; Calderón, F.; Benito, S. The Influence of Non-*Saccharomyces* Species on Wine Fermentation Quality Parameters. *Fermentation* **2019**, *5*, 54. [CrossRef]

89. Berbegal, C.; Spano, G.; Fragasso, M.; Grieco, F.; Russo, P.; Capozzi, V. Starter cultures as biocontrol strategy to prevent *Brettanomyces bruxellensis* proliferation in wine. *Appl. Microbiol. Biotechnol.* **2018**, *102*, 569–576. [CrossRef]

90. Di Toro, M.R.; Capozzi, V.; Beneduce, L.; Alexandre, H.; Tristezza, M.; Durante, M.; Tufariello, M.; Grieco, F.; Spano, G. Intraspecific biodiversity and 'spoilage potential' of *Brettanomyces bruxellensis* in Apulian wines. *LWT Food Sci. Technol.* **2015**, *60*, 102–108. [CrossRef]

91. Russo, P.; Fragasso, M.; Berbegal, C.; Grieco, F.; Spano, G.; Capozzi, V. Microorganisms Able to Produce Biogenic Amines and Factors Affecting Their Activity. In *Biogenic Amines in Food*; RCS Publishing: Cambridge, UK, 2019; pp. 18–40.

92. Russo, P.; Capozzi, V.; Spano, G.; Corbo, M.R.; Sinigaglia, M.; Bevilacqua, A. Metabolites of Microbial Origin with an Impact on Health: Ochratoxin A and Biogenic Amines. *Front. Microbiol.* **2016**, *7*, 482. [CrossRef] [PubMed]

93. Berbegal, C.; Borruso, L.; Fragasso, M.; Tufariello, M.; Russo, P.; Brusetti, L.; Spano, G.; Capozzi, V. A Metagenomic-Based Approach for the Characterization of Bacterial Diversity Associated with Spontaneous Malolactic Fermentations in Wine. *Int. J. Mol. Sci.* **2019**, *20*, 3980. [CrossRef] [PubMed]

94. Zhang, S.; Chen, X.; Zhong, Q.; Zhuang, X.; Bai, Z. Microbial Community Analyses Associated with Nine Varieties of Wine Grape Carposphere Based on High-Throughput Sequencing. *Microorganisms* **2019**, *7*, 668. [CrossRef] [PubMed]

95. Garofalo, C.; Berbegal, C.; Grieco, F.; Tufariello, M.; Spano, G.; Capozzi, V. Selection of indigenous yeast strains for the production of sparkling wines from native Apulian grape varieties. *Int. J. Food Microbiol.* **2018**, *285*, 7–17. [CrossRef]

96. Gómez Brandón, M.; Aira, M.; Kolbe, A.R.; de Andrade, N.; Pérez-Losada, M.; Domínguez, J. Rapid Bacterial Community Changes during Vermicomposting of Grape Marc Derived from Red Winemaking. *Microorganisms* **2019**, *7*, 473. [CrossRef] [PubMed]

97. Ab Rahman, S.F.; Singh, E.; Pieterse, C.M.J.; Schenk, P.M. Emerging microbial biocontrol strategies for plant pathogens. *Plant Sci. Int. J. Exp. Plant Biol.* **2018**, *267*, 102–111. [CrossRef]

98. Bleve, G.; Grieco, F.; Cozzi, G.; Logrieco, A.; Visconti, A. Isolation of epiphytic yeasts with potential for biocontrol of *Aspergillus carbonarius* and *A. niger* on grape. *Int. J. Food Microbiol.* **2006**, *108*, 204–209. [CrossRef]

99. Berbegal, C.; Garofalo, C.; Russo, P.; Pati, S.; Capozzi, V.; Spano, G. Use of Autochthonous Yeasts and Bacteria in Order to Control *Brettanomyces bruxellensis* in Wine. *Fermentation* **2017**, *3*, 65. [CrossRef]

100. Nally, M.C.; Pesce, V.M.; Maturano, Y.P.; Assaf, L.R.; Toro, M.E.; de Figueroa, L.C.; Vazquez, F. Antifungal modes of action of *Saccharomyces* and other biocontrol yeasts against fungi isolated from sour and grey rots. *Int. J. Food Microbiol.* **2015**, *204*, 91–100. [CrossRef] [PubMed]

101. Pretscher, J.; Fischkal, T.; Branscheidt, S.; Jäger, L.; Kahl, S.; Schlander, M.; Thines, E.; Claus, H. Yeasts from Different Habitats and Their Potential as Biocontrol Agents. *Fermentation* **2018**, *4*, 31. [CrossRef]

102. López-Seijas, J.; García-Fraga, B.; da Silva, A.F.; Sieiro, C. Wine Lactic Acid Bacteria with Antimicrobial Activity as Potential Biocontrol Agents against *Fusarium oxysporum* f. sp. lycopersici. *Agronomy* **2020**, *10*, 31. [CrossRef]

103. Leyva Salas, M.; Mounier, J.; Valence, F.; Coton, M.; Thierry, A.; Coton, E. Antifungal Microbial Agents for Food Biopreservation—A Review. *Microorganisms* **2017**, *5*, 37. [CrossRef]

104. Vargas, M.; Garrido, F.; Zapata, N.; Tapia, M. Isolation and selection of epiphytic yeast for biocontrol of *Botrytis cinerea* pers. on table grapes. *Chil. J. Agric. Res.* **2012**, *72*, 332. [CrossRef]

105. Parafati, L.; Vitale, A.; Restuccia, C.; Cirvilleri, G. Biocontrol ability and action mechanism of food-isolated yeast strains against *Botrytis cinerea* causing post-harvest bunch rot of table grape. *Food Microbiol.* **2015**, *47*, 85–92. [CrossRef]

106. Parafati, L.; Vitale, A.; Polizzi, G.; Restuccia, C.; Cirvilleri, G. Understanding the mechanism of biological control of postharvest phytopathogenic moulds promoted by food isolated yeasts. *Acta Hortic.* **2016**, 93–100. [CrossRef]

107. Kasfi, K.; Taheri, P.; Jafarpour, B.; Tarighi, S. Identification of epiphytic yeasts and bacteria with potential for biocontrol of grey mold disease on table grapes caused by *Botrytis cinerea*. *Span. J. Agric. Res.* **2018**, *16*, 23. [CrossRef]

108. Nally, M.C.; Pesce, V.M.; Maturano, Y.P.; Muñoz, C.J.; Combina, M.; Toro, M.E.; de Figueroa, L.I.C.; Vazquez, F. Biocontrol of Botrytis cinerea in table grapes by non-pathogenic indigenous *Saccharomyces cerevisiae* yeasts isolated from viticultural environments in Argentina. *Postharvest Biol. Technol.* **2012**, *64*, 40–48. [CrossRef]

109. Cordero-Bueso, G.; Mangieri, N.; Maghradze, D.; Foschino, R.; Valdetara, F.; Cantoral, J.M.; Vigentini, I. Wild Grape-Associated Yeasts as Promising Biocontrol Agents against Vitis vinifera Fungal Pathogens. *Front. Microbiol.* **2017**, *8*. [CrossRef]

110. Lemos Junior, W.J.F.; Bovo, B.; Nadai, C.; Crosato, G.; Carlot, M.; Favaron, F.; Giacomini, A.; Corich, V. Biocontrol Ability and Action Mechanism of *Starmerella bacillaris* (Synonym *Candida zemplinina*) Isolated from Wine Musts against Gray Mold Disease Agent *Botrytis cinerea* on Grape and Their Effects on Alcoholic Fermentation. *Front. Microbiol.* **2016**, *7*. [CrossRef] [PubMed]

111. Qin, X.; Xiao, H.; Xue, C.; Yu, Z.; Yang, R.; Cai, Z.; Si, L. Biocontrol of gray mold in grapes with the yeast *Hanseniaspora uvarum* alone and in combination with salicylic acid or sodium bicarbonate. *Postharvest Biol. Technol.* **2015**, *100*, 160–167. [CrossRef]

112. Mewa-Ngongang, M.; Du Plessis, H.W.; Ntwampe, S.K.O.; Chidi, B.S.; Hutchinson, U.F.; Mekuto, L.; Jolly, N.P. Fungistatic and Fungicidal Properties of *Candida pyralidae* Y1117, *Pichia kluyveri* Y1125 and *Pichia kluyveri* Y1164 on the Biocontrol of Botrytis Cinereal. In Proceedings of the 10th International Conference on Advances in Science, Engineering, Technology and Healthcare (ASETH-18), Cape Town, South Africa, 19–20 November 2018; pp. 19–20.

113. Mewa-Ngongang, M.; Du Plessis, H.W.; Ntwampe, S.K.O.; Chidi, B.S.; Hutchinson, U.F.; Mekuto, L.; Jolly, N.P. The Use of *Candida pyralidae* and *Pichia kluyveri* to Control Spoilage Microorganisms of Raw Fruits Used for Beverage Production. *Foods* **2019**, *8*, 454. [CrossRef] [PubMed]

114. Zhou, Q.; Fu, M.; Xu, M.; Chen, X.; Qiu, J.; Wang, F.; Yan, R.; Wang, J.; Zhao, S.; Xin, X. Application of antagonist *Bacillus amyloliquefaciens* NCPSJ7 against Botrytis cinerea in postharvest Red Globe grapes. *Food Sci. Nutr.* **2020**, *8*, 1499–1508. [CrossRef] [PubMed]

115. Chen, X.; Wang, Y.; Gao, Y.; Gao, T.; Zhang, D. Inhibitory Abilities of *Bacillus* Isolates and Their Culture Filtrates against the Gray Mold Caused by *Botrytis cinerea* on Postharvest Fruit. *Plant Pathol. J.* **2019**, *35*, 425–436. [CrossRef]

116. Passera, A.; Venturini, G.; Battelli, G.; Casati, P.; Penaca, F.; Quaglino, F.; Bianco, P.A. Competition assays revealed *Paenibacillus pasadenensis* strain R16 as a novel antifungal agent. *Microbiol. Res.* **2017**, *198*, 16–26. [CrossRef]

117. Walker, G.M.; Stewart, G.G. *Saccharomyces cerevisiae* in the Production of Fermented Beverages. *Beverages* **2016**, *2*, 30. [CrossRef]

118. Albertin, W.; Marullo, P.; Aigle, M.; Dillmann, C.; de Vienne, D.; Bely, M.; Sicard, D. Population Size Drives Industrial *Saccharomyces cerevisiae* Alcoholic Fermentation and Is under Genetic Control. *Appl. Environ. Microbiol.* **2011**, *77*, 2772–2784. [CrossRef]

119. Feyder, S.; De Craene, J.-O.; Bär, S.; Bertazzi, D.L.; Friant, S. Membrane Trafficking in the Yeast *Saccharomyces cerevisiae* Model. *Int. J. Mol. Sci.* **2015**, *16*, 1509–1525. [CrossRef]

120. Tristezza, M.; Tufariello, M.; Capozzi, V.; Spano, G.; Mita, G.; Grieco, F. The Oenological Potential of *Hanseniaspora uvarum* in Simultaneous and Sequential Co-fermentation with *Saccharomyces cerevisiae* for Industrial Wine Production. *Front. Microbiol.* **2016**, *7*. [CrossRef]

121. Capozzi, V.; Berbegal, C.; Tufariello, M.; Grieco, F.; Spano, G. Impact of co-inoculation of *Saccharomyces cerevisiae*, *Hanseniaspora uvarum* and *Oenococcus oeni* autochthonous strains in controlled multi starter grape must fermentations. *LWT* **2019**, *109*, 241–249. [CrossRef]

122. Liu, H.M.; Guo, J.H.; Cheng, Y.J.; Luo, L.; Liu, P.; Wang, B.Q.; Deng, B.X.; Long, C.A. Control of gray mold of grape by *Hanseniaspora uvarum* and its effects on postharvest quality parameters. *Ann. Microbiol.* **2010**, *60*, 31–35. [CrossRef]

123. Liu, H.M.; Guo, J.H.; Luo, L.; Liu, P.; Wang, B.Q.; Cheng, Y.J.; Deng, B.X.; Long, C.A. Improvement of *Hanseniaspora uvarum* biocontrol activity against gray mold by the addition of ammonium molybdate and the possible mechanisms involved. *Crop Prot.* **2010**, *29*, 277–282. [CrossRef]

124. Russo, P.; Englezos, V.; Capozzi, V.; Pollon, M.; Río Segade, S.; Rantsiou, K.; Spano, G.; Cocolin, L. Effect of mixed fermentations with *Starmerella bacillaris* and *Saccharomyces cerevisiae* on management of malolactic fermentation. *Food Res. Int.* **2020**, *134*, 109246. [CrossRef] [PubMed]

125. Russo, P.; Tufariello, M.; Renna, R.; Tristezza, M.; Taurino, M.; Palombi, L.; Capozzi, V.; Rizzello, C.G.; Grieco, F. New Insights into the Oenological Significance of *Candida zemplinina*: Impact of Selected Autochthonous Strains on the Volatile Profile of Apulian Wines. *Microorganisms* **2020**, *8*, 628. [CrossRef]

126. Masneuf-Pomarede, I.; Juquin, E.; Miot-Sertier, C.; Renault, P.; Laizet, Y.; Salin, F.; Alexandre, H.; Capozzi, V.; Cocolin, L.; Colonna-Ceccaldi, B.; et al. The yeast *Starmerella bacillaris* (synonym *Candida zemplinina*) shows high genetic diversity in winemaking environments. *FEMS Yeast Res.* **2015**, *15*. [CrossRef]

127. Mateus, D.; Sousa, S.; Coimbra, C.; Rogerson, F.S.; Simões, J. Identification and Characterization of Non-*Saccharomyces* Species Isolated from Port Wine Spontaneous Fermentations. *Foods* **2020**, *9*, 120. [CrossRef]

128. Saravanakumar, D.; Ciavorella, A.; Spadaro, D.; Garibaldi, A.; Gullino, M.L. *Metschnikowia pulcherrima* strain MACH1 outcompetes *Botrytis cinerea*, *Alternaria alternata* and *Penicillium expansum* in apples through iron depletion. *Postharvest Biol. Technol.* **2008**, *49*, 121–128. [CrossRef]

129. Roudil, L.; Russo, P.; Berbegal, C.; Albertin, W.; Spano, G.; Capozzi, V. Non-*Saccharomyces* Commercial Starter Cultures: Scientific Trends, Recent Patents and Innovation in the Wine Sector. *Recent Pat. Food Nutr. Agric.* **2019**, *11*. [CrossRef]

130. Compant, S.; Duffy, B.; Nowak, J.; Clément, C.; Barka, E.A. Use of plant growth-promoting bacteria for biocontrol of plant diseases: Principles, mechanisms of action, and future prospects. *Appl. Environ. Microbiol.* **2005**, *71*, 4951–4959. [CrossRef]

131. Khan, N.; Maymon, M.; Hirsch, A.M. Combating *Fusarium* Infection Using Bacillus-Based Antimicrobials. *Microorganisms* **2017**, *5*, 75. [CrossRef] [PubMed]

132. Khan, N.; Martínez-Hidalgo, P.; Ice, T.A.; Maymon, M.; Humm, E.A.; Nejat, N.; Sanders, E.R.; Kaplan, D.; Hirsch, A.M. Antifungal Activity of *Bacillus* Species Against *Fusarium* and Analysis of the Potential Mechanisms Used in Biocontrol. *Front. Microbiol.* **2018**, *9*. [CrossRef] [PubMed]

133. Campa-Siqueiros, P.; Vallejo-Cohen, S.; Corrales-Maldonado, C.; Martínez-Téllez, M.Á.; Vargas-Arispuro, I.; Ávila-Quezada, G. Reduction in the incidence of grey mold in table grapes due to the volatile effect of a garlic extract. *Rev. Mex. Fitopatol. Mex. J. Phytopathol.* **2017**, *35*. [CrossRef]

134. Zhang, Z.; Qin, G.; Li, B.; Tian, S. Effect of Cinnamic Acid for Controlling Gray Mold on Table Grape and Its Possible Mechanisms of Action. *Curr. Microbiol.* **2015**, *71*, 396–402. [CrossRef]

135. Wang, Y.; Liu, X.; Chen, T.; Xu, Y.; Tian, S. Antifungal effects of hinokitiol on development of *Botrytis cinerea* in vitro and in vivo. *Postharvest Biol. Technol.* **2020**, *159*, 111038. [CrossRef]

136. Xueuan, R.; Dandan, S.; Zhuo, L.; Qingjun, K. Effect of mint oil against *Botrytis cinerea* on table grapes and its possible mechanism of action. *Eur. J. Plant Pathol.* **2018**, *151*, 321–328. [CrossRef]

137. Jiang, L.; Jin, P.; Wang, L.; Yu, X.; Wang, H.; Zheng, Y. Methyl jasmonate primes defense responses against *Botrytis cinerea* and reduces disease development in harvested table grapes. *Sci. Hortic.* **2015**, *192*, 218–223. [CrossRef]

138. Xu, D.; Deng, Y.; Xi, P.; Yu, G.; Wang, Q.; Zeng, Q.; Jiang, Z.; Gao, L. Fulvic acid-induced disease resistance to *Botrytis cinerea* in table grapes may be mediated by regulating phenylpropanoid metabolism. *Food Chem.* **2019**, *286*, 226–233. [CrossRef]

139. Xu, D.; Deng, Y.; Han, T.; Jiang, L.; Xi, P.; Wang, Q.; Jiang, Z.; Gao, L. In vitro and in vivo effectiveness of phenolic compounds for the control of postharvest gray mold of table grapes. *Postharvest Biol. Technol.* **2018**, *139*, 106–114. [CrossRef]

140. Youssef, K.; de Oliveira, A.G.; Tischer, C.A.; Hussain, I.; Roberto, S.R. Synergistic effect of a novel chitosan/silica nanocomposites-based formulation against gray mold of table grapes and its possible mode of action. *Int. J. Biol. Macromol.* **2019**, *141*, 247–258. [CrossRef]

141. Kanetis, L.; Exarchou, V.; Charalambous, Z.; Goulas, V. Edible coating composed of chitosan and *Salvia fruticosa* Mill. extract for the control of grey mould of table grapes. *J. Sci. Food Agric.* **2017**, *97*, 452–460. [CrossRef] [PubMed]

142. Guerra, I.C.D.; De Oliveira, P.D.L.; Santos, M.M.F.; Lúcio, A.S.S.C.; Tavares, J.F.; Barbosa-Filho, J.M.; Madruga, M.S.; De Souza, E.L. The effects of composite coatings containing chitosan and Mentha (*piperita* L. or *x villosa* Huds) essential oil on postharvest mold occurrence and quality of table grape cv. Isabella. *Innov. Food Sci. Emerg. Technol.* **2016**, *34*, 112–121. [CrossRef]

143. Takma, D.K.; Korel, F. Impact of preharvest and postharvest alginate treatments enriched with vanillin on postharvest decay, biochemical properties, quality and sensory attributes of table grapes. *Food Chem.* **2017**, *221*, 187–195. [CrossRef] [PubMed]

144. Sivakumar, D.; Bautista-Baños, S. A review on the use of essential oils for postharvest decay control and maintenance of fruit quality during storage. *Crop Prot.* **2014**, *64*, 27–37. [CrossRef]

145. Zhou, R.; Mo, Y.; Li, Y.; Zhao, Y.; Zhang, G.; Hu, Y. Quality and internal characteristics of Huanghua pears (*Pyrus pyrifolia* Nakai, cv. Huanghua) treated with different kinds of coatings during storage. *Postharvest Biol. Technol.* **2008**, *49*, 171–179. [CrossRef]

146. Azarakhsh, N.; Osman, A.; Ghazali, H.M.; Tan, C.P.; Mohd Adzahan, N. Lemongrass essential oil incorporated into alginate-based edible coating for shelf-life extension and quality retention of fresh-cut pineapple. *Postharvest Biol. Technol.* **2014**, *88*, 1–7. [CrossRef]

147. Sharma, R.R.; Singh, D.; Singh, R. Biological control of postharvest diseases of fruits and vegetables by microbial antagonists: A review. *Biol. Control* **2009**, *50*, 205–221. [CrossRef]

148. Underhill, S.J.R.; Joshua, L.; Zhou, Y. A Preliminary Assessment of Horticultural Postharvest Market Loss in the Solomon Islands. *Horticulturae* **2019**, *5*, 5. [CrossRef]

149. Kumar, D.; Kalita, P. Reducing Postharvest Losses during Storage of Grain Crops to Strengthen Food Security in Developing Countries. *Foods* **2017**, *6*, 8. [CrossRef]

150. McKenzie, T.J.; Singh-Peterson, L.; Underhill, S.J.R. Quantifying Postharvest Loss and the Implication of Market-Based Decisions: A Case Study of Two Commercial Domestic Tomato Supply Chains in Queensland, Australia. *Horticulturae* **2017**, *3*, 44. [CrossRef]

151. Mahajan, P.V.; Caleb, O.J.; Singh, Z.; Watkins, C.B.; Geyer, M. Postharvest treatments of fresh produce. *Philos. Transact. R. Soc. A Math. Phys. Eng. Sci.* **2014**, *372*. [CrossRef] [PubMed]

152. Romanazzi, G.; Smilanick, J.L.; Feliziani, E.; Droby, S. Integrated management of postharvest gray mold on fruit crops. *Postharvest Biol. Technol.* **2016**, *113*, 69–76. [CrossRef]

153. Shi, Z.; Deng, J.; Wang, F.; Liu, Y.; Jiao, J.; Wang, L.; Zhang, J. Individual and combined effects of bamboo vinegar and peach gum on postharvest grey mould caused by *Botrytis cinerea* in blueberry. *Postharvest Biol. Technol.* **2019**, *155*, 86–93. [CrossRef]

154. Panebianco, S.; Vitale, A.; Polizzi, G.; Scala, F.; Cirvilleri, G. Enhanced control of postharvest citrus fruit decay by means of the combined use of compatible biocontrol agents. *Biol. Control* **2015**, *84*, 19–27. [CrossRef]

155. Zhang, X.; Min, D.; Li, F.; Ji, N.; Meng, D.; Li, L. Synergistic effects of L-arginine and methyl salicylate on alleviating postharvest disease caused by *Botrysis cinerea* in tomato fruit. *J. Agric. Food Chem.* **2017**, *65*, 4890–4896. [CrossRef]

156. Liguori, G.; D'Aquino, S.; Sortino, G.; De Pasquale, C.; Inglese, P. Effects of passive and active modified atmosphere packaging conditions on quality parameters of minimally processed table grapes during cold storage. *J. Berry Res.* **2015**, *5*, 131–143. [CrossRef]

157. Taghavi, T.; Kim, C.; Rahemi, A. Role of Natural Volatiles and Essential Oils in Extending Shelf Life and Controlling Postharvest Microorganisms of Small Fruits. *Microorganisms* **2018**, *6*, 104. [CrossRef]

158. Chiabrando, V.; Garavaglia, L.; Giacalone, G. The Postharvest Quality of Fresh Sweet Cherries and Strawberries with an Active Packaging System. *Foods* **2019**, *8*, 335. [CrossRef]

159. Chang, X.; Lu, Y.; Li, Q.; Lin, Z.; Qiu, J.; Peng, C.; Brennan, C.S.; Guo, X. The Combination of Hot Air and Chitosan Treatments on Phytochemical Changes during Postharvest Storage of 'Sanhua' Plum Fruits. *Foods* **2019**, *8*, 338. [CrossRef]

160. Sharma, S.; Pareek, S.; Sagar, N.A.; Valero, D.; Serrano, M. Modulatory Effects of Exogenously Applied Polyamines on Postharvest Physiology, Antioxidant System and Shelf Life of Fruits: A Review. *Int. J. Mol. Sci.* **2017**, *18*, 1789. [CrossRef]

161. Camele, I.; Altieri, L.; De Martino, L.; De Feo, V.; Mancini, E.; Rana, G.L. In Vitro Control of Post-Harvest Fruit Rot Fungi by Some Plant Essential Oil Components. *Int. J. Mol. Sci.* **2012**, *13*, 2290–2300. [CrossRef] [PubMed]

162. Zhang, H.; Li, R.; Liu, W. Effects of Chitin and Its Derivative Chitosan on Postharvest Decay of Fruits: A Review. *Int. J. Mol. Sci.* **2011**, *12*, 917–934. [CrossRef] [PubMed]

163. Mamoci, E.; Cavoski, I.; Simeone, V.; Mondelli, D.; Al-Bitar, L.; Caboni, P. Chemical Composition and In Vitro Activity of Plant Extracts from Ferula communis and Dittrichia viscosa against Postharvest Fungi. *Molecules* **2011**, *16*, 2609–2625. [CrossRef]

164. Ghosh, R.; Barman, S.; Mukhopadhyay, A.; Mandal, N.C. Biological control of fruit-rot of jackfruit by rhizobacteria and food grade lactic acid bacteria. *Biol. Control* **2015**, *83*, 29–36. [CrossRef]

165. Zhimo, V.Y.; Biasi, A.; Kumar, A.; Feygenberg, O.; Salim, S.; Vero, S.; Wisniewski, M.; Droby, S. Yeasts and Bacterial Consortia from Kefir Grains Are Effective Biocontrol Agents of Postharvest Diseases of Fruits. *Microorganisms* **2020**, *8*, 428. [CrossRef] [PubMed]

166. Iglesias, M.B.; Abadias, M.; Anguera, M.; Sabata, J.; Viñas, I. Antagonistic effect of probiotic bacteria against foodborne pathogens on fresh-cut pear. *LWT—Food Sci. Technol.* **2017**, *81*, 243–249. [CrossRef]

167. Grieco, F.; Castellano, M.A.; Di Sansebastiano, G.P.; Maggipinto, G.; Neuhaus, J.-M.; Martelli, G.P. Subcellular localization and in vivo identification of the putative movement protein of olive latent virus 2. *J. Gen. Virol.* **1999**, *80*, 1103–1109. [CrossRef]

168. Russo, P.; Peña, N.; de Chiara, M.L.V.; Amodio, M.L.; Colelli, G.; Spano, G. Probiotic lactic acid bacteria for the production of multifunctional fresh-cut cantaloupe. *Food Res. Int.* **2015**, *77*, 762–772. [CrossRef]

169. Russo, P.; de Chiara, M.L.V.; Vernile, A.; Amodio, M.L.; Arena, M.P.; Capozzi, V.; Massa, S.; Spano, G. Fresh-Cut Pineapple as a New Carrier of Probiotic Lactic Acid Bacteria. *BioMed Res. Int.* **2014**, *2014*, 1–9. [CrossRef]

170. Arena, M.P.; Capozzi, V.; Russo, P.; Drider, D.; Spano, G.; Fiocco, D. Immunobiosis and probiosis: Antimicrobial activity of lactic acid bacteria with a focus on their antiviral and antifungal properties. *Appl. Microbiol. Biotechnol.* **2018**, *102*, 9949–9958. [CrossRef]

© 2020 by the authors. Licensee MDPI, Basel, Switzerland. This article is an open access article distributed under the terms and conditions of the Creative Commons Attribution (CC BY) license (http://creativecommons.org/licenses/by/4.0/).

Review

Plasma-Activated Water (PAW) as a Disinfection Technology for Bacterial Inactivation with a Focus on Fruit and Vegetables

Aswathi Soni [1,*], Jonghyun Choi [2] and Gale Brightwell [1,3]

[1] Food Assurance, AgResearch, Palmerston North 4442, New Zealand; Gale.Brightwell@agresearch.co.nz
[2] The New Zealand Institute for Plant and Food Research Ltd., Private Bag 3230, Waikato Mail Centre, Hamilton 3240, New Zealand; Jonghyun.Choi@plantandfood.co.nz
[3] New Zealand Food Safety Science Research Centre, Palmerston North 4474, New Zealand
[*] Correspondence: Aswathi.Soni@agresearch.co.nz; Tel.: +64-21-0860-7979

Abstract: Plasma-activated water (PAW) is generated by treating water with cold atmospheric plasma (CAP) using controllable parameters, such as plasma-forming voltage, carrier gas, temperature, pulses, or frequency as required. PAW is reported to have lower pH, higher conductivity, and higher oxygen reduction potential when compared with untreated water due to the presence of reactive species. PAW has received significant attention from researchers over the last decade due to its non-thermal and non-toxic mode of action especially for bacterial inactivation. The objective of the current review is to develop a summary of the effect of PAW on bacterial strains in foods as well as model systems such as buffers, with a specific focus on fruit and vegetables. The review elaborated the properties of PAW, the effect of various treatment parameters on its efficiency in bacterial inactivation along with its usage as a standalone technology as well as a hurdle approach with mild thermal treatments. A section highlighting different models that can be employed to generate PAW alongside a direct comparison of the PAW characteristics on the inactivation potential and the existing research gaps are also included. The mechanism of action of PAW on the bacterial cells and any reported effects on the sensory qualities and shelf life of food has been evaluated. Based on the literature, it can be concluded that PAW offers a significant potential as a non-chemical and non-thermal intervention for bacterial inactivation, especially on food. However, the applicability and usage of PAW depend on the effect of environmental and bacterial strain-based conditions and cost-effectiveness.

Keywords: cold atmospheric plasma; microbes; disinfection; non-hazardous; inactivation; foodborne pathogen

check for **updates**

Citation: Soni, A.; Choi, J.; Brightwell, G. Plasma-Activated Water (PAW) as a Disinfection Technology for Bacterial Inactivation with a Focus on Fruit and Vegetables. *Foods* **2021**, *10*, 166. https://doi.org/10.3390/foods10010166

Received: 18 December 2020
Accepted: 12 January 2021
Published: 15 January 2021

Publisher's Note: MDPI stays neutral with regard to jurisdictional claims in published maps and institutional affiliations.

Copyright: © 2021 by the authors. Licensee MDPI, Basel, Switzerland. This article is an open access article distributed under the terms and conditions of the Creative Commons Attribution (CC BY) license (https://creativecommons.org/licenses/by/4.0/).

1. Introduction

Food spoilage is defined as a change in any food product that leads to a significant reduction in its sensory qualities, such as color, texture, and overall smell, due to physical damage or chemical changes (e.g., oxidation), and thus rendering it unacceptable by the consumer [1]. These changes are mainly the result of microbial growth and metabolism in the food, which may lead to the production of enzymes that facilitate to reactions resulting in deleterious by-products affecting the food. These by-products vary in different types of food and can lead to adverse sensory properties, including the presence of slime, off-odors, and off-flavors. Bacterial strains associated with spoilage include *Pectobacterium carotovorum* [2,3], *Brochothrix thermosphacta* [4], *Clostridium perfringens* [5], *Bacillus spp.* [6,7], *Pseudomonas fragi* [8], *Pseudomonas fluorescens* [9], *Shewanella putrefaciens* [10], *Serratia liquefaciens* [11] and *Hafnia alvei* [12]. Food spoilage is a primary concern for food industries due to susceptible loss of shelf life and hence the economic losses followed by a long-term impact on consumer preferences. Nevertheless, food spoilage is also a threat to the environment as it leads to excessive wastage that end ups in the landfill, which does not contribute to sustainable living. This is supported by a survey conducted in 2018, which indicated

Foods **2021**, *10*, 166. https://doi.org/10.3390/foods10010166

that 30–50% of the total food produced exclusively by one country in a year ends up in the landfill, with the contribution from households, processing industries, food services, primary production sector and retails being 53%, 19%, 12%, 11%, 5%, respectively [13]. Minimizing food spoilage by employing multiple interventions might help not only the food industries but also the environment.

Another concern for the food processing industries and the regulatory authorities is food poisoning due to bacterial growth in food. Some bacterial strains are capable of producing toxins under certain conditions either in the food itself or inside the human body once live bacterial cells are ingested, while others are enteropathogenic and entero-invasive pathogens [14]. Few examples of foodborne pathogens include *Clostridium botulinum* [15,16], *B. cereus* [17], *Staphylococcus aureus* [18,19], *Listeria monocytogenes* [20], *Salmonella enterica* serovar typhimurium [21], *Salmonella* spp. [22], *E. coli* O157:H7 [23], *C. perfringens* [24], *Shigella* spp. [25], *Yersina* spp. [26] and *Campylobacter jejuni* [27]. Food poisoning has been a major public health concern, particularly regarding outbreaks affecting immunocompromised individuals and infants and thus may lead to adverse social and economic effects. Although many food-poisoning cases go under-reported due to quick recovery and almost minimum effect on healthy individuals, some still have adverse effects on immunocompromised individuals and infants [28]. Alternate disinfection technologies that do not employ thermal treatments or harmful chemicals could be valuable options for minimizing contamination and growth of microbial contaminants leading to either spoilage or food poisoning. Such sustainable technologies are of great importance to the vast ever-growing population with increasing demand of food across the globe.

Recently, the application of advanced oxidation processes (AOPs) for the decontamination of fruit and vegetables has been widely investigated. These technologies include electrolyzed water [29,30], gaseous ozone [3,31], UV light [32,33], and cold plasma [33,34]. One of these oxidation technologies is plasma-activated water (PAW), which is generated by treating water with cold atmospheric plasma (CAP) using controllable parameters such as plasma-forming voltage, carrier gas, temperature, pulses, or frequency as required. Plasma has been recognized as the fourth state of matter. It is the ionized gas usually produced when gas molecules are exposed to the electric field, forming reactive species and ions [35]. PAW has received significant attention from researchers over the last decade due to its non-thermal and non-toxic mode of action, which is mainly due to the reactive species that could react with the bacterial structural components and later organelles, leading to death [36]. This review summarizes the role of PAW as a disinfectant for bacterial decontamination on the surface of foods with a focus on vegetables and fruit [36–40]. Although most of the studies have investigated the combined effect of PAW with other interventions such as heat, their standalone effect has also been reported.

2. Systems for PAW Generation

The fundamental method of generation of PAW involves operating a plasma generator inside the water to generate the ions, which lead to reactive species for bacterial inactivation. There are various combinations and models in the literature leading to difference in the final outputs; these are outlined in Table 1.

Table 1. The effect of different generation conditions and the characteristics of PAW.

Gas and Additional Features	Gap (Between Water Surface and the Upper Electrode)	AC Voltage and Frequency	Quantitative Changes after Generation Time	References
Grounded copper electrode (diameter 0.5 mM) on top and a capillary tube to generate bubbles	10 mM	3–6 kV, 3–10 kHz	pH changed from 6.75 to 3.77 and NO_2^- concentration changed from 3.77 µM to 8.686 µM in 15 min of activation	[41]
Plasma jet unit coaxial tungsten electrode and a quartz tube (diameter of 700 µM)	nil	6–10 kV, 7.0 kHz	Not determined	[42]
Plasma jet with RD1004 rotating nozzle	8.1-cM	voltage (295 V), air pressure (1990 mBar), and frequency (22.5 kHz)	pH changed from 6.5 to 3.1 and Oxidation-reduction potential (ORP) increased from 376.54 to 534.52 RmV in 5 min	[43]
Atmospheric pressure plasma jet (patent atmospheric pressure plasma jet (APPJ))	0 mM	3.0 kV and 16 kHz	pH reduced from 7 to 3.2 in 20 min and the ORP increased from 310 to 510 mV.	[40]
1. DC-driven streamer corona. 2. Transient spark discharge	10 mM	(~10 mA) with a 5–20 kHz repetition rate, 10 kV	The pH changed reduced by 4 units.	[44]
Air plasma generator with copper electrodes and quartz dielectric	2 cM	20 kHz, high voltage (not specified)	The pH changed from 6.8 to 2.3, ORP changed from 250 to 540 mV.	[45]
Atmospheric cold plasma jet	7.5 cM	20 kHz, 30 kV	The pH changed from 5.88 to 2.85, ORP changed from 406.1 to 565.40 mV.	[46]

Most studies have indicated an immediate drop in pH and an increase in electrical conductivity and the ORP as a result of the formation of reactive species in the PAW samples (Table 1). However, the increase of change in these properties cannot be directly correlated with a single factor or reason. When PAW is produced, the gaseous species from either the working or the atmospheric gas enters the liquid-gas interface and as a result there are complex reactions leading to the non-equilibrium, hence generation of the ionic moieties [44,47,48]. This process is highly influenced by the electric field and also using bubble implosions which hence the movement as well as dispersion of the phenomenon across the interface [49]. A recent review suggests that the electrical breakdown in water can occur without a phase change such as evaporating liquid and condensing or dissolving vapor [48]. The factors affecting the changes in PAW during activation may depend on multiple factors. For example, increase in discharge power, which is a direct function of applied voltage, would affect the increase in electric conductivity of the PAW [47]. On the other hand, in another study Vlad et al. showed that increase in treatment time would increase bacterial inactivation by PAW [50]. Although these studies reported above (Table 1) have used different set ups for the PAW generation, it could still be concluded that the efficiency can be a combined effect of two or more factors such as PAW activation time, temperature, power used and the aeration or bubbling to improve the formation of reactive oxygen species (ROS) [51].

Physicochemical Properties of PAW

PAW shows lower pH, higher conductivity and higher oxygen reduction potential when compared with untreated water [44,52]. The reduction in pH is due to the formation of acidic chemical species, which result in a steep decrease from pH 7 to pH 3 within 5–10 min of activation, but with little change thereafter [37,53]. Oxidation-reduction potential (ORP) can be defined as the ability of any solution to acquire or loose electrons to an electrode, and this property of PAW is much more prominent as compared with non-activated water. ORP of PAW depends on the strength of activation, which further depends on the applied voltage, carrier gas and other parameters leading to an increase of up to 63% [54]. Conductivity is the ability of any solution to allow current to pass through it and is reported to significantly increase due to plasma activation, primarily because of the generation of ions [45]. With a plasma jet that was operated from a 10 kHz sinusoidal high-voltage power source with 18 kV peak-to-peak AC voltage using pre-mixed oxygen and argon, the pH reduced from 7 to 3, ORP increased from 250 to 550 mV, conductivity rose from 0 to 410 µS/cm and temperature increased from 25 to 30 °C after 15 min of activation [55]. With a similar plasma source, when PAW was produced using 0.40–0.42 kV AC voltage, the conductivity increased from 5 to 20 mS/cm, the ORP value increased from 180 to 250 mV, pH decreased from 7.0 to 6.0, and the temperature increased from 20 to 40 °C [56]. Hence, the change in the reactive species of PAW are measurable as ORP and pH, and these changes directly show their effect on the potential to attack and disrupt the bacterial membranes during inactivation [40]. These effects are explained in detail in this review.

3. The Effect of PAW on Bacterial Inactivation

3.1. The Effect of PAW on Vegetables and Fruit

PAW has been known to show inactivation in the model systems such as water and buffer, indicating potential to be used in liquid food matrix. However, different food matrices can have different effects on the microbial adhesion and resistance, which could alter their efficiency in several ways. Table 2 summarizes the reports on microbial inactivation specifically on surfaces of vegetables and fruit. In most cases described here, the use of PAW is as a washing intervention resulting in inactivation of up to 3 $\log_{10}$ CFU/mL of different strains (Table 2).

Table 2. The effect of PAW on microbial inactivation on fruit and vegetables.

Bacterial Strain and Food	PAW Discharger	Input AC Voltage (kV)	Power (W)	Plasma Inducement Time (min)	Treatment Time (min)	Inactivation Achieved	Ref
Aerobic bacteria on apple slices	Hollow fiber-based cold micro plasma jet	10 kV	-	10	5	2 $\log_{10}$ CFU/mL	[42]
Aerobic bacterial counts on mung bean sprouts	Atmospheric pressure plasma jet (APPJ) system based on gliding arc discharge in air	5 kV	750 W	0.5	30	2.3 $\log_{10}$ CFU/g	[57]
S. aureus on fresh-cut kiwifruit slices	Micro plasma array	-	-	-	30	1.8 $\log_{10}$ CFU/g	[58]
S. aureus on strawberries	APPJ	18 kV		20	15	2.0 $\log_{10}$ CFU/g	[55]
Total bacterial count on baby spinach leaves	Surface barrier discharge (SBD) reactor	11 kV	36 W	20	2	1 $\log_{10}$ CFU/g	[59]
Saccharomyces cerevisiae CICC 1374 inoculated on grape berries	Plasma jet with electrodes and a dielectric	8.2 kV	-	60	30	0.5 $\log_{10}$ CFU/g	[60]

The efficiency of PAW cannot be directly compared with that of heat treatment such as thermal pasteurization which can lead to at least 6 $\log_{10}$ CFU/mL of non-spore-forming bacteria depending on the food matrix. However, PAW treatment has been reported to be effective in increasing the shelf life of fresh food.

PAW has also shown positive results in inhibiting the growth of bacteria, molds, and yeasts in fresh-cut apples in a study by Liu et al. [42]. In this study, fresh-cut apple slices were immersed in PAW generated at 8 kV for 5 min at room temperature and the treated slices were stored at 4 ± 1 °C and 90% relative humidity (RH) for 12 days. Aerobic bacteria count reduced by 1.05 $\log_{10}$ CFU/g on day 12 of refrigerated storage [42]. Additionally, on day 12, a reduction equivalent to 0.64, and 1.04 $\log_{10}$ CFU/g was observed in molds, and yeasts, respectively. PAW treated apples showed absence in total aerobic bacteria for the first two days as compared with the 3 $\log_{10}$ CFU on the control samples which were immersed in distilled water and stored under the same conditions as PAW treated slices. After 12 days, the total aerobic count reached up to 4–5 $\log_{10}$ CFU/mL in PAW treated slices which were still less than the untreated sample by 2 $\log_{10}$ CFU/mL [42]. Yeast and mold numbers were also found to be lesser by 0.5 and 1 $\log_{10}$ CFU/g in the PAW treated samples as compared with the control. The delayed onset of bacterial growth and hence increase in numbers in the PAW treated samples were postulated as a result of sub-lethal injury due to the reactive species. A similar effect on shelf life using PAW as a disinfecting agent was observed on strawberries, where PAW treated (10 kHz sinusoidal AC voltage source at 18 kV) strawberries showed less than 1 $\log_{10}$ CFU/g inactivation of *S. aureus*. Complete absence of any growth of hyphae of filamentous fungi after six days of storage at 20 ± 2 °C and 70 ± 5% RH indicated a potential of PAW to counter fungal spoilage [55]. PAW has been reported to not only reduce the total aerobic counts but also reduced the ability of growth by fungi and yeasts, which indicates that PAW has the ability to alter the microbial community of the food-based ecosystem, which needs further investigation [57]. Interestingly, fungal growth has been reported to be significantly reduced by PAW treatment on strawberries by Ma et al. [55]. The rough and uneven surfaces of strawberries act as physical barriers to protect microflora from ethanol and UV light inactivation. However, PAW was could access these areas, indicating a potential of PAW over other interventions.

3.2. The Effect of PAW on Microbial Inactivation in Model Systems

Buffers, diluents and salt solutions are often used as the dispersion matrix in place of real food for microbial inactivation assays. This not only helps to minimize the effects intrinsic characteristics and composition on the results, but also provides opportunity to modify the intrinsic factors according to specific requirements. The studies specifically reported on inactivation of non-spore-forming bacterial cells in model systems using PAW are listed in Table 3.

Table 3. The effect of PAW on microbial inactivation in non-food matrix.

Bacterial Strain and Model System	PAW Discharger	Input Voltage (kV)	Current, Power, or Frequency	Plasma Inducement Time (mins)	Treatment Time (mins)	Inactivation Achieved ($\log_{10}$ CFU/mL)	Ref
Enterobacter aerogenes in water	Plasma jet with rotating nozzle	295 V	22.5 kHz	NR	5	1.92 ± 0.70	[43]
Enterobacter aerogenes in plasma-activated acidified buffer (PAAB)	Plasma jet with rotating nozzle	295 V	22.5 kHz	NR	5	5.11 ± 0.63	[43]
Hafnia alvei in disinfecting PAW solutions (9.9 mL)	Glidarc system	NR	NR	5	30	>5.0	[61]
Staphylococcus epidermidis ATCC 12228,	Gliding arc type	NR	NR	5	15	<6	[62]
Leuconostoc mesenteroides CNRZ 1468 cells in tryptic soy broth (TSB)	Gliding arc type	NR	NR	5	20	<6	[62]
H. alvei CIP 5731 cells in TSB	Gliding arc type	NR	NR	5	20	<6	[62]
S. putrefaciens in sterile phosphate-buffered saline (PBS)	Atmospheric cold plasma jet	22 kV	NR	5	30	>7	[46]
S. aureus (NCTC-8325) cells in TSB	Corona discharge	10 kV	5 mA 50 mW	30	60	5.52 ± 0.23	[63]
E. coli in sterile PBS	Atmospheric cold plasma jet	30 kV	NR	5	60	5.7	[46]

NR = not reported.

Bacterial spores are resistant to thermal treatment (based on the decimal reduction time or D-value), several chemicals, and dehydration. In most of the cases, either moderate to high thermal treatment or a combined effect of multiple inactivation technologies need to be applied for spore inactivation in food [64]. PAW has been investigated as an alternative to thermal treatments or harmful chemicals against bacterial spores. For example, a study by Bai et al. [65] showed that up to 2 $\log_{10}$ CFU/mL reduction in the population of *B. cereus* spores could be achieved by a single treatment of PAW at 55 °C (input power of 650 W for 60 s). This study also indicated that the inactivation kinetics fitted the log-logistic model, which assumes there could be differences in the inactivation kinetics of any bacterial population. As per this model, the whole bacterial spore population can be divided into subpopulations. The first subpopulation has spores that are entirely resistant to the treatment. The second has the ones that can repair themselves post-treatment and the third are fully vulnerable to the treatment [66]. According to Bai et al. [65], the inactivation was enhanced by reducing the volume of activation from 100 to 50 mL and by lowering the spore inoculum from 10^7 to 10^5 CFU/mL.

The effects of PAW on biofilms have also been explored. A study by Xu et al. [63] indicated that both the plasma inducement time and the plasma treatment time could affect the bacterial cells in their ability to generate biofilms. Apart from more than 5 $\log_{10}$ CFU/unit reduction of *S. aureus* cells in the biofilm, the regrowth capacity of the surviving cells was reduced by 30% as compared with the untreated cells [63]. This effect is postulated as a result of regulation of genes (*SarA, IcaA, SigB, Rbf, LuxS*) involved in biofilm formation, especially in response to disinfectants such as hydrogen peroxide. [63,67]. On the contrary, in a study by Charoux et al. it was seen that PAW as a standalone method was ineffective in inactivating *Escherichia coli* K12 cells in biofilms; however the synergistic effects of airborne acoustic ultrasound and PAW enhanced the inactivation potential by 2 log CFU/mL [68]. A study by Hozak et al. also indicated that PAW was ineffective in inactivation of Gram-positive and Gram-negative cells in the biofilm [69]. Similar results were obtained when PAW was found to be more efficient in inactivating *S. epidermidis*, *L. mesenteroides*, and *H. alvei* in planktonic forms as compared with adherent forms in biofilms [62]. Postulated reasons for the reduced effect by PAW against the bacterial cells in biofilm could be due to the inability to overcome the resistance of the bacterial cells in stationary phase unlike those from overnight culture (planktonic cells) in exponential phase. Alternatively, the short-lived reactive species, which are effective against the planktonic cells might not be effective in the dense consortium of the biofilm, which consists of extra cellular matric as well as well-organized clumps of bacterial cells. Although the mechanism of resistance of bacterial biofilms is yet to be investigated in detail, it can be concluded that PAW would need to be combined with at least one more technology to enhance its efficiency in cleaning regimes for biofilm removal.

3.3. The Effect of PAW as a Hurdle Intervention against Bacterial Inactivation

PAW has also been investigated as a hurdle technology or a combined synergistic approach against bacterial inactivation, especially in case of bacterial spores. The combination of PAW at 40 °C and 55 °C for the decontamination of *B. cereus* spores resulted in an inactivation equivalent to 1.54 and 2.12 $\log_{10}$ CFU/g of *B. cereus* spores, respectively after 60 min of exposure [70]. The detailed analysis on the structure of the treated *B. cereus* spores using transmission electron microscopy, scanning electron microscopy and the propidium iodide (PI) assays indicated visible disruption of external structure (spore coat and cortex) along with leakage of intracellular contents [70]. PI is a fluorochrome capable of binding and labelling DNA fragments, only when there is some level of damage to an otherwise intact cell membrane, which leads to permeability of the PI into the cell [71].

B. cereus spores consist of multi-layered protective structure including a spore coat and a peptidoglycan cortex [72], which acts as a barrier to the external chemical disinfectants, including phenols, organic acids, quaternary ammonium compounds (QACs) biguanides and alcohols [73,74]. Images of spores treated with PAW showed specific disruption or damage on the external layers, which include the spore coat made of peptidoglycan layers. The structural damage is indicative of oxidative stress and mild thermal stress [70]. Ultrasound technology (40 Hz, 220 W) when combined with PAW at 40 °C for 60 min has been tested against *S. aureus* and *E. coli*. Ultrasound technology employs mechanical sound waves of low to high frequencies (20 kHz–1.5 MHz) to have multiple applications including bacterial inactivation, by imparting a damage to the cell walls as the mechanism of action [75]. In combination with PAW, this mechanism against cell structural integrity could be enhanced. As a result, inactivation equivalent to 1.33 $\log_{10}$ CFU/mL of *E. coli* K12 and 0.83 $\log_{10}$ CFU/mL of *S. aureus* was achieved, as compared with less than 0.5 $\log_{10}$ CFU/mL for both strains under the same conditions with PAW alone [36]. There was no significant difference among the pH, EC, and ORP [36] between the PAW samples treated with and without ultrasonication, indicating that any change in the parameters of acidity, EC and ORP was due to the presence of reactive species generated by plasma activation and not by ultrasonication. The PAW treatment used in this study was for a set time (up to 60 min) while being held at three different temperatures (4, 25 and 40 °C), followed by ultrasonication at 40 Hz and output power of 220 W. This study demonstrated that ultrasound increased the penetration of the reactive species generated by PAW into the bacterial cells, and though the inactivation was significantly different by <1 $\log_{10}$ CFU/mL when PAW treatment with and without ultrasound was compared, the scanning electron microscopic images showed increased porosity of the PAW and ultrasound treated samples as compared with the samples treated with de-ionized water [36]. PAW and mild heat (50 °C) for 6 min has been reported to be effective against *S. cerevisiae* leading to an inactivation equivalent to 4.4 $\log_{10}$ CFU/mL as compared with a relatively lower inactivation equivalent to 0.27 and 1.92 $\log_{10}$ CFU/mL with PAW at 25 °C and mild heat at 50 °C for 6 min as standalone treatments [76]. This study further investigated the damage caused on the cells due to treatment using scanning electron microscopy, which revealed the appearance of visible deformation on cell surface along with complete distortion of parts of the cell wall in contrast to the images showing cell integrity post-treatment by PAW alone and mild heat alone. The PI permeability assay also supported these findings where the intensity of penetration of dye increased significantly ($p < 0.05$, by 132.14-fold), indicating compromised cytoplasmic membranes and thereby increased binding of the dye to cellular DNA and RNA [76,77]. PAW and mild heat (50, 52.5, and 55 °C, 30 min) has been reported to inactivate *Saccharomyces cerevisiae* by 2.4 $\log_{10}$ CFU/g when inoculated on grapes [78]. *S. cerevisiae* has been reported in spoilage of fresh fruit and fruit juices [79]. Therefore, these studies on successful inactivation of yeast indicate a potential application of PAW towards extension of shelf life with minimum effect on the quality attributes.

In another study, the combined effect of PAW (18 kV for 120 min) and mild heat at 60 °C led to inactivation equivalent to 3.4 and 3.7 $\log_{10}$ CFU/g of *L. monocytogenes* and *S. aureus*, respectively, on salted Chinese cabbages [38]. The treatment was performed in a sequential approach where initial PAW treatment was performed for 10 min, followed by heating in a water bath (60 °C) for 5 min. Interestingly, this treatment of PAW (18 kV for 120 min) and mild heat at 60 °C also resulted in 4, 5.7, 4.0, and 2.6 $\log_{10}$ CFU/mL reduction in mesophilic aerobic counts, lactic acid bacteria, yeast, and molds and coliforms, respectively, which were all present in the background microflora of the untreated samples [38]. This study indicated a potential application of PAW with mild heat as a disinfection approach in Korean market for cabbage kimchi [38] which is preferably cleaned using a non-thermal approach to minimize changes in the sensory attributes for a fresh-like quality. Considering the studies reported, it can be concluded that PAW has demonstrated efficiency as a hurdle approach either in combination or sequentially to eliminate the bacteria or yeast from food surfaces. However, the impact on the formation of the reactive species in the food needs further research as a case-by-case basis.

3.4. PAW-Mechanism of Action on Bacterial Cells

The mechanism of disinfection using PAW can be explained as the oxidative stress on the cell membranes of the bacterial cells. The cell membrane is disrupted, followed by further damages to organelles, proteins and nucleic acid leading to the cell death [45,61]. The inactivation of bacterial cells by PAW has been reported as an effect of the reactive species on the bacterial cell membranes and the organelles. The reactive species can interact with and hence damage the bacterial cell membranes through lipid peroxidation, which disrupts the structure followed by leakage, morphology change, DNA damage and disruptions of the functional structures of the proteins [80]. There is a clear agreement in the literature on the dominant reactive species responsible for the mechanism of action against bacterial cells. For example, hydroxyl radicals, ozone, nitric radicals (NO_2^-), and nitrate ions (NO_3^-) are considered capable of interacting with the biological components of the cell membrane and later the cytoplasmic components and organelles inside the bacterial cell. The process of structural disruption leads to functional damage and physiological imbalance. For example, three reactive species (NO_3^-, H_2O_2, O_3) generated in PAW using a dielectric barrier discharge (DBD)-atmospheric cold plasma (ACP) system showed an effect in inhibiting the growth of spoilage microbes as measured through total viable counts on shrimps [81]. In this case, the total microbial numbers increased significantly ($p < 0.05$) from 3.9 to 8.6 $\log_{10}$ CFU/g during nine days of storage in control samples that were washed using tap water, whereas the growth of microbes were slower when washed with PAW. The increase was only from 3.9 to 4.4 $\log_{10}$ CFU/g for the initial 6 days, and reached 6.5 log10 CFU/g on the 9th day when stored at 5 °C, indicating possible sub-lethal stress on the initial population, which might have hindered the growth rate [81]. This study did not establish the individual effect of each of the reactive species. However, in general, reactive species are known to attack intracellular superoxide dismutase (SOD), catalase (CAT), peroxidase, and oxidases, which impart the form the defense mechanism of the bacterial cells against stress [82,83]. This initial resistance is overcome when the concentration of reactive species increases above the capacity of in-built defense systems, leading to commencement of the oxidative damage involving lipids, DNA, and proteins. This process is defined as oxidative stress, which can lead to cell death [84,85]. Short-lived reactive species such as oxygen radicals lead to the immediate lethal effect on the bacterial cells, whereas long-lived species such as nitrate ions, hydrogen peroxide radicals and ozone might lead to a reduction in the recovery and growth of the sub-lethally injured cells [81]. The reduction in pH was only a difference of 0.2 units and in this case, the reduction in microbial growth rate cannot be attributed to decrease in pH balance outside and hence inside the cells [81]. Additionally, the gas used for the creation of the PAW does not seem to indicate a direct influence on the inactivation of bacterial cells [81]. However, this needs further investigation before being established as a fact.

4. The Effect of PAW on Sensory Attributes and Consumer Acceptance

Consumer demand for fresh-like, minimally processed or natural food has been a growing area of interest for food scientists and industries. These trends call for technologies that only result in insignificant or very mild changes due to physical, and chemical reactions that affect the "fresh-like quality" [86–88]. PAW is a chemical-free and non-thermal intervention and has the potential towards achieving the minimal processing target. However, there are no sensory studies using trained panels reported in the literature. The changes associated with quality and appearance, which also contribute to likability, have only been investigated at the laboratory scale. For example, Chinese bayberries treated by PAW showed a significantly ($p < 0.05$) higher firmness (measured using a texture analyzer) by 1.2 newton (N) when stored at 3 °C for eight days as compared with untreated berries [89]. The PAW treatment used in this study was a hollow electrode dielectric barrier structure (HEDBS) with air as the working gas at a flow rate of 260 L/h using an AC power source with a frequency of 20 kHz. The treatment time in this study was 0.5 min at room temperature (25 °C) [89]. In red grapes, samples treated with PAW had a higher rate of color change (red or violet to dark violet) than untreated controls [89], as assessed using three parameters of color index for red grapes (CIRG), namely lightness (L), hue angle (H) and chroma (C), over eight days of storage at 3 °C [90]. The change in color was attributed to the impact of ROS on anthocyanins [89]. However, in response to this oxidative stress, these berries were found to produce more antioxidants and accelerate the antioxidant enzymes to scavenge excessive ROS, which also related to the higher CIRG value as compared to the control [89].

In a similar study, the effect of PAW (11 kV for 20 min) on the color (L, a, and b values) of spinach leaves after storage at 4 °C for eight days was compared with the changes in the untreated samples [59]. The control (untreated) and PAW treated samples showed no significant difference in color evaluated using L, a, and b values, which indicated the "fresh-like appearance". It was shown that multiple rinsing with water affected the color as there was an increase in the yellowness (b values), which was not seen with the PAW treated samples [59]. In another study by Ma et al. [55] using PAW treatment (10 kHz sinusoidal high-voltage source at 18 kV), the change in color and firmness of the treated and untreated strawberries were compared after storage at 20 ± 2 °C and $70 \pm 5\%$ RH for four days [55]. No significant difference in color (L, a, and b values) and firmness (N) was observed between the treated and untreated strawberries, which showed that PAW did not induce any change in these attributes and hence would resemble fresh-like qualities. However, the firmness of the treated and the untreated samples was reduced after storage when compared with the results obtained on day zero, indicating a natural phenomenon of loss of the turgor pressure post-harvesting, especially when held at temperatures above refrigeration [55]. Color is considered to be a very influential parameter of consumer's perception of food's healthiness and purchase intention. Hence, the reports on the color changes are indicative of consumer preference and perception [91]. The absence of any significant change on color would indicate minimum effect of PAW on fruit and vegetables and can be perceived as "minimally processed".

5. Future Implications, Research Gap, and Conclusions

PAW has promising potentials to be used as a non-chemical decontamination agent for food. Although no harmful chemicals are used in the generation of PAW, the presence of residual reactive species in food during the shelf life is an area of investigation that remains underexplored. At the same time, the cost and energy required to generate efficiency against bacterial contaminants are high. The research gap on the specific effect of microbial inactivation and the impact of specific intrinsic and extrinsic food properties opens opportunities for comprehensive research and application. Therefore, PAW could emerge as a non-hazardous, non-chemical disinfectant for food industries which could be used as a hurdle or as a standalone step depending on the microbial population being targeted. The successful and safe application of PAW in food-grade systems would also depend on the optimization of analytical methods for detecting reactive species with minimum interference due to the composition of food being tested.

Most of the work done on PAW as a disinfectant is on food (fruit and vegetables) surface in lab settings where the controlled environment helps to identify variables and mechanism of action. There is a significant gap in the literature on the use of PAW as a disinfectant in real-life supply chain scenarios (e.g., packaging house and cool store). At the same time, the specific compositional changes due to any residual reactive species needs to be studied throughout shelf life to ensure that the structural changes of the food (if any) do not affect any sensory attributes. The effect of PAW on bacterial spores food and model systems need an elaborative investigation. Also, it is important to confirm that the reactive species in PAW do not lead to the germination onset in spores as this could adversely affect the shelf life or safety of food. Alternatively, if PAW can induce germination, it has the potential to be used as a hurdle before thermal inactivation thereby reducing the heat exposure and energy required to inactivate bacterial spores. There is no evidence in the literature to compare the effect of PAW of the decimal reduction time (D-value) of spores and vegetative bacterial cells, highlighting the research gap on whether this resistance (D-value) is comparable to the well-established thermal resistance, which is much higher in spores. Similarly, there is a lack of direct comparison of the resistance exhibited by Gram-positive and Gram-negative vegetative cells. Future work in these areas would enhance the knowledge of PAW disinfection in food processing.

Author Contributions: Conceptualization, A.S. and G.B.; investigation, A.S.; writing—original draft preparation, A.S.; writing—review and editing, J.C., A.S., and G.B.; project administration, G.B.; funding acquisition, G.B. All authors have read and agreed to the published version of the manuscript.

Funding: This work is funded through the AgResearch Strategic Science Investment Fund (SSIF) Food Integrity (contract A25768).

Acknowledgments: The research outlined in this study was supported by the AgResearch Ltd. Strategic Science Investment Fund (SSIF). The authors wish to thank John Mills and Shuyan Wu for their technical inputs and Joshua Hadi for the help with proofreading.

Conflicts of Interest: The authors declare no conflict of interest.

References

1. Gram, L.; Ravn, L.; Rasch, M.; Bruhn, J.B.; Christensen, A.B.; Givskov, M. Food spoilage—interactions between food spoilage bacteria. *Int. J. Food Microbiol.* **2002**, *78*, 79–97. [CrossRef]
2. Hassenberg, K.; Frohling, A.; Geyer, M.; Schluter, O.; Herppich, W. Ozonated wash water for inhibition of Pectobacterium carotovorum on carrots and the effect on the physiological behaviour of produce. *Eur. J. Hortic. Sci.* **2008**, *73*, 37.
3. Pinto, L.; Yaseen, T.; Caputo, L.; Furiani, C.; Carboni, C.; Baruzzi, F. Application of passive refrigeration and gaseous ozone to reduce postharvest losses on red chicory. In Proceedings of the VI International Conference Postharvest Unlimited 1256, Madrid, Spain, 8 November 2019; pp. 419–426.
4. Greer, G.G.; Dilts, B.D. Control of Brochothrix thermosphacta spoilage of pork adipose tissue using bacteriophages. *J. Food Prot.* **2002**, *65*, 861–863. [CrossRef] [PubMed]
5. De Jong, J. Spoilage of an acid food product by *Clostridium perfringens*, *C. barati* and *C. butyricum*. *Int. J. Food Microbiol.* **1989**, *8*, 121–132. [CrossRef]
6. Thompson, J.M.; Dodd, C.E.R.; Waites, W.M. Spoilage of bread by bacillus. *Int. Biodeterior. Biodegrad.* **1993**, *32*, 55–66. [CrossRef]
7. André, S.; Vallaeys, T.; Planchon, S. Spore-forming bacteria responsible for food spoilage. *Res. Microbiol.* **2017**, *168*, 379–387. [CrossRef] [PubMed]
8. Miller, A.; Scanlan, R.; Lee, J.; Libbey, L. Identification of the volatile compounds produced in sterile fish muscle (Sebastes melanops) by Pseudomonas fragi. *Appl. Microbiol.* **1973**, *25*, 952–955. [CrossRef]
9. Kumar, H.; Franzetti, L.; Kaushal, A.; Kumar, D. Pseudomonas fluorescens: A potential food spoiler and challenges and advances in its detection. *Ann. Microbiol.* **2019**, *69*, 873–883. [CrossRef]
10. Jørgensen, B.R.; Huss, H.H. Growth and activity of Shewanella putrefaciens isolated from spoiling fish. *Int. J. Food Microbiol.* **1989**, *9*, 51–62. [CrossRef]
11. Salgado, C.A.; Baglinière, F.; Vanetti, M.C.D. Spoilage potential of a heat-stable lipase produced by Serratia liquefaciens isolated from cold raw milk. *LWT* **2020**, *126*, 109289. [CrossRef]
12. Hanna, M.; Smith, G.; Hall, L.; Vanderzant, C. Role of Hafnia alvei and a Lactobacillus species in the spoilage of vacuum-packaged strip loin steaks. *J. Food Prot.* **1979**, *42*, 569–571. [CrossRef] [PubMed]
13. Racz, A.; Vasiljev Marchesi, V.; Crnković, I. Economical, environmental and ethical impact of food wastage in hospitality and other global industries. *JAHR* **2018**, *9*, 25–42. [CrossRef]
14. Reis, R.S.d.; Horn, F. Enteropathogenic Escherichia coli, Samonella, Shigella and Yersinia: Cellular aspects of host-bacteria interactions in enteric diseases. *Gut Pathog.* **2010**, *2*, 8. [CrossRef] [PubMed]
15. Crane, J.K. Preformed Bacterial Toxins. *Clin. Lab. Med.* **1999**, *19*, 583–599. [CrossRef]
16. Hennekinne, J.-A.; De Buyser, M.-L.; Dragacci, S. Staphylococcus aureus and its food poisoning toxins: Characterization and outbreak investigation. *Fems Microbiol. Rev.* **2012**, *36*, 815–836. [CrossRef]
17. Ceuppens, S.; Rajkovic, A.; Heyndrickx, M.; Tsilia, V.; Van De Wiele, T.; Boon, N.; Uyttendaele, M. Regulation of toxin production by Bacillus cereus and its food safety implications. *Crit. Rev. Microbiol.* **2011**, *37*, 188–213. [CrossRef]
18. Halpin-Dohnalek, M.I.; Marth, E.H. Staphylococcus aureus: Production of extracellular compounds and behavior in foods-A review. *J. Food Prot.* **1989**, *52*, 267–282. [CrossRef]
19. Le Loir, Y.; Baron, F.; Gautier, M. Staphylococcus aureus and food poisoning. *Genet. Mol. Res.* **2003**, *2*, 63–76.
20. Farber, J.; Peterkin, P. Listeria monocytogenes, a food-borne pathogen. *Microbiol. Mol. Biol. Rev.* **1991**, *55*, 476–511. [CrossRef]
21. Davies, A.; O'neill, P.; Towers, L.; Cooke, M. An outbreak of Salmonella typhimurium DT104 food poisoning associated with eating beef. *Commun. Dis. Rep. Cdr. Rev.* **1996**, *6*, R159.
22. Smith, S.; Alao, F.; Goodluck, H.; Fowora, M.; Bamidele, M.; Omonigbehin, E.; Coker, A. Prevalence of Salmonella typhi among food handlers from bukkas in Nigeria. *Br. J. Biomed. Sci.* **2008**, *65*, 158–160. [CrossRef] [PubMed]
23. Doyle, M.P. Escherichia coli O157: H7 and its significance in foods. *Int. J. Food Microbiol.* **1991**, *12*, 289–301. [CrossRef]
24. García, S.; Heredia, N. Clostridium perfringens: A dynamic foodborne pathogen. *Food. Bioprocess Technol.* **2011**, *4*, 624–630. [CrossRef]

25. Warren, B.; Parish, M.; Schneider, K. Shigella as a foodborne pathogen and current methods for detection in food. *Crit. Rev. Food Sci. Nutr.* **2006**, *46*, 551–567. [CrossRef]

26. Zadernowska, A.; Chajęcka-Wierzchowska, W.; Łaniewska-Trokenheim, Ł. Yersinia enterocolitica: A dangerous, but often ignored, foodborne pathogen. *Food Rev. Int.* **2014**, *30*, 53–70. [CrossRef]

27. Altekruse, S.F.; Stern, N.J.; Fields, P.I.; Swerdlow, D.L. Campylobacter jejuni—An emerging foodborne pathogen. *Emerg. Infect. Dis.* **1999**, *5*, 28. [CrossRef]

28. Al-Kharousi, Z.S.; Guizani, N.; Al-Sadi, A.M.; Al-Bulushi, I.M.; Shaharoona, B. Hiding in fresh fruits and vegetables: Opportunistic pathogens may cross geographical barriers. *Int. J. Microbiol.* **2016**, *2016*. [CrossRef]

29. Pinto, L.; Ippolito, A.; Baruzzi, F. Control of spoiler Pseudomonas spp. on fresh cut vegetables by neutral electrolyzed water. *Food Microbiol.* **2015**, *50*, 102–108. [CrossRef]

30. Pinto, L.; Baruzzi, F.; Ippolito, A. Recent advances to control spoilage microorganisms in washing water of fruits and vegetables: The use of electrolyzed water. In Proceedings of the III International Symposium on Postharvest Pathology: Using Science to Increase Food Availability 1144, Bari, Italy, 7 November 2016; pp. 379–384.

31. Pinto, L.; Caputo, L.; Quintieri, L.; de Candia, S.; Baruzzi, F. Efficacy of gaseous ozone to counteract postharvest table grape sour rot. *Food Microbiol.* **2017**, *66*, 190–198. [CrossRef]

32. Fan, X.; Huang, R.; Chen, H. Application of ultraviolet C technology for surface decontamination of fresh produce. *Trends Food Sci. Technol.* **2017**, *70*, 9–19. [CrossRef]

33. Pinto, L.; Baruzzi, F.; Cocolin, L.; Malfeito-Ferreira, M. Emerging technologies to control Brettanomyces spp. in wine: Recent advances and future trends. *Trends Food Sci. Technol.* **2020**, *99*, 88–100. [CrossRef]

34. Wang, R.; Nian, W.; Wu, H.; Feng, H.; Zhang, K.; Zhang, J.; Zhu, W.; Becker, K.; Fang, J. Atmospheric-pressure cold plasma treatment of contaminated fresh fruit and vegetable slices: Inactivation and physiochemical properties evaluation. *Eur. Phys. J. D* **2012**, *66*, 276. [CrossRef]

35. Lackmann, J.-W.; Bandow, J.E. Inactivation of microbes and macromolecules by atmospheric-pressure plasma jets. *Appl. Microbiol. Biotechnol.* **2014**, *98*, 6205–6213. [CrossRef] [PubMed]

36. Royintarat, T.; Choi, E.H.; Boonyawan, D.; Seesuriyachan, P.; Wattanutchariya, W. Chemical-free and synergistic interaction of ultrasound combined with plasma-activated water (PAW) to enhance microbial inactivation in chicken meat and skin. *Sci. Rep.* **2020**, *10*, 1–14. [CrossRef]

37. Thirumdas, R.; Kothakota, A.; Annapure, U.; Siliveru, K.; Blundell, R.; Gatt, R.; Valdramidis, V.P. Plasma activated water (PAW): Chemistry, physico-chemical properties, applications in food and agriculture. *Trends Food Sci. Technol.* **2018**, *77*, 21–31. [CrossRef]

38. Choi, E.J.; Park, H.W.; Kim, S.B.; Ryu, S.; Lim, J.; Hong, E.J.; Byeon, Y.S.; Chun, H.H. Sequential application of plasma-activated water and mild heating improves microbiological quality of ready-to-use shredded salted kimchi cabbage (*Brassica pekinensis* L.). *Food Control* **2019**, *98*, 501–509. [CrossRef]

39. Xu, Y.; Tian, Y.; Ma, R.; Liu, Q.; Zhang, J. Effect of plasma activated water on the postharvest quality of button mushrooms, Agaricus bisporus. *Food Chem.* **2016**, *197*, 436–444. [CrossRef]

40. Lin, C.-M.; Chu, Y.-C.; Hsiao, C.-P.; Wu, J.-S.; Hsieh, C.-W.; Hou, C.-Y. The optimization of plasma-activated water treatments to inactivate Salmonella enteritidis (ATCC 13076) on shell eggs. *Foods* **2019**, *8*, 520. [CrossRef]

41. Sajib, S.A.; Billah, M.; Mahmud, S.; Miah, M.; Hossain, F.; Omar, F.B.; Roy, N.C.; Hoque, K.M.F.; Talukder, M.R.; Kabir, A.H.; et al. Plasma activated water: The next generation eco-friendly stimulant for enhancing plant seed germination, vigor and increased enzyme activity, a study on black gram (*Vigna mungo* L.). *Plasma Chem. Plasma Process.* **2020**, *40*, 119–143. [CrossRef]

42. Liu, C.; Chen, C.; Jiang, A.; Sun, X.; Guan, Q.; Hu, W. Effects of plasma-activated water on microbial growth and storage quality of fresh-cut apple. *Innov. Food Sci. Emerg. Technol.* **2020**, *59*, 102256. [CrossRef]

43. Joshi, I.; Salvi, D.; Schaffner, D.W.; Karwe, M.V. Characterization of Microbial Inactivation Using Plasma-Activated Water and Plasma-Activated Acidified Buffer. *J. Food Prot.* **2018**, *81*, 1472–1480. [CrossRef] [PubMed]

44. Machala, Z.; Tarabová, B.; Sersenová, D.; Janda, M.; Hensel, K. Chemical and antibacterial effects of plasma activated water: Correlation with gaseous and aqueous reactive oxygen and nitrogen species, plasma sources and air flow conditions. *J Phys. D* **2018**, *52*, 034002. [CrossRef]

45. Shen, J.; Tian, Y.; Li, Y.; Ma, R.; Zhang, Q.; Zhang, J.; Fang, J. Bactericidal Effects against S. aureus and Physicochemical Properties of Plasma Activated Water stored at different temperatures. *Sci. Rep.* **2016**, *6*, 28505. [CrossRef] [PubMed]

46. Zhao, Y.-M.; Ojha, S.; Burgess, C.M.; Sun, D.-W.; Tiwari, B.K. Inactivation efficacy and mechanisms of plasma activated water on bacteria in planktonic state. *J. Appl. Microbiol.* **2020**, *129*, 1248–1260. [CrossRef] [PubMed]

47. Anderson, C.E.; Cha, N.R.; Lindsay, A.D.; Clark, D.S.; Graves, D.B. The Role of Interfacial Reactions in Determining Plasma–Liquid Chemistry. *Plasma Chem. Plasma Process.* **2016**, *36*, 1393–1415. [CrossRef]

48. Bruggeman, P.J.; Kushner, M.J.; Locke, B.R.; Gardeniers, J.G.E.; Graham, W.G.; Graves, D.B.; Hofman-Caris, R.C.H.M.; Maric, D.; Reid, J.P.; Ceriani, E.; et al. Plasma-liquid interactions: A review and roadmap. *Plasma Sources Sci. Technol.* **2016**, *25*, 053002. [CrossRef]

49. Zhao, Y.-M.; Patange, A.; Sun, D.-W.; Tiwari, B. Plasma-activated water: Physicochemical properties, microbial inactivation mechanisms, factors influencing antimicrobial effectiveness, and applications in the food industry. *Compr. Rev. Food Sci. Food Saf.* **2020**, *19*, 3951–3979. [CrossRef]

50. Vlad, I.; Martin, C.; Toth, A.; Papp, J.; Anghel, S. Bacterial inhibition effect of plasma activated water. *Rom. Rep. Phys.* **2019**, *71*, 602.
51. Zhou, R.; Zhou, R.; Wang, P.; Xian, Y.; Mai-Prochnow, A.; Lu, X.; Cullen, P.; Ostrikov, K.K.; Bazaka, K. Plasma-activated water: Generation, origin of reactive species and biological applications. *J. Phys. D* **2020**, *53*, 303001. [CrossRef]
52. Lu, P.; Boehm, D.; Bourke, P.; Cullen, P.J. Achieving reactive species specificity within plasma-activated water through selective generation using air spark and glow discharges. *Plasma Process. Polym.* **2017**, *14*, 1600207. [CrossRef]
53. Thirumdas, R.; Sarangapani, C.; Annapure, U.S. Cold Plasma: A novel Non-Thermal Technology for Food Processing. *Food Biophys.* **2015**, *10*, 1–11. [CrossRef]
54. Zhang, Q.; Ma, R.; Tian, Y.; Su, B.; Wang, K.; Yu, S.; Zhang, J.; Fang, J. Sterilization Efficiency of a Novel Electrochemical Disinfectant against Staphylococcus aureus. *Environ. Sci. Technol.* **2016**, *50*, 3184–3192. [CrossRef]
55. Ma, R.; Wang, G.; Tian, Y.; Wang, K.; Zhang, J.; Fang, J. Non-thermal plasma-activated water inactivation of food-borne pathogen on fresh produce. *J. Hazard. Mater.* **2015**, *300*, 643–651. [CrossRef] [PubMed]
56. Tian, Y.; Ma, R.; Zhang, Q.; Feng, H.; Liang, Y.; Zhang, J.; Fang, J. Assessment of the Physicochemical Properties and Biological Effects of Water Activated by Non-thermal Plasma Above and Beneath the Water Surface. *Plasma Process. Polym.* **2015**, *12*, 439–449. [CrossRef]
57. Xiang, Q.; Liu, X.; Liu, S.; Ma, Y.; Xu, C.; Bai, Y. Effect of plasma-activated water on microbial quality and physicochemical characteristics of mung bean sprouts. *Innov. Food Sci. Emerg. Technol.* **2019**, *52*, 49–56. [CrossRef]
58. Zhao, Y.; Chen, R.; Liu, D.; Wang, W.; Niu, J.; Xia, Y.; Qi, Z.; Zhao, Z.; Song, Y. Effect of Nonthermal Plasma-Activated Water on Quality and Antioxidant Activity of Fresh-Cut Kiwifruit. *IEEE Trans. Plasma Sci.* **2019**, *47*, 4811–4817. [CrossRef]
59. Risa Vaka, M.; Sone, I.; García Álvarez, R.; Walsh, J.L.; Prabhu, L.; Sivertsvik, M.; Noriega Fernández, E. Towards the Next-Generation Disinfectant: Composition, Storability and Preservation Potential of Plasma Activated Water on Baby Spinach Leaves. *Foods* **2019**, *8*, 692. [CrossRef]
60. Guo, J.; Huang, K.; Wang, X.; Lyu, C.; Yang, N.; Li, Y.; Wang, J. Inactivation of Yeast on Grapes by Plasma-Activated Water and Its Effects on Quality Attributes. *J. Food Prot.* **2017**, *80*, 225–230. [CrossRef]
61. Naïtali, M.; Kamgang-Youbi, G.; Herry, J.-M.; Bellon-Fontaine, M.-N.; Brisset, J.-L. Combined Effects of Long-Living Chemical Species during Microbial Inactivation Using Atmospheric Plasma-Treated Water. *Appl. Environ. Microbiol.* **2010**, *76*, 7662–7664. [CrossRef]
62. Kamgang-Youbi, G.; Herry, J.-M.; Meylheuc, T.; Brisset, J.-L.; Bellon-Fontaine, M.-N.; Doubla, A.; Naïtali, M. Microbial inactivation using plasma-activated water obtained by gliding electric discharges. *Lett. Appl. Microbiol.* **2009**, *48*, 13–18. [CrossRef]
63. Xu, Z.; Zhou, X.; Yang, W.; Zhang, Y.; Ye, Z.; Hu, S.; Ye, C.; Li, Y.; Lan, Y.; Shen, J.; et al. In vitro antimicrobial effects and mechanism of air plasma-activated water on Staphylococcus aureus biofilm. *Plasma Process. Polym.* **2020**, *17*, 1900270. [CrossRef]
64. Soni, A.; Oey, I.; Silcock, P.; Bremer, P. Bacillus Spores in the Food Industry: A Review on Resistance and Response to Novel Inactivation Technologies. *Compr. Rev. Food Sci. Food Saf.* **2016**, *15*, 1139–1148. [CrossRef] [PubMed]
65. Bai, Y.; Idris Muhammad, A.; Hu, Y.; Koseki, S.; Liao, X.; Chen, S.; Ye, X.; Liu, D.; Ding, T. Inactivation kinetics of Bacillus cereus spores by Plasma activated water (PAW). *Food Res. Int.* **2020**, *131*, 109041. [CrossRef] [PubMed]
66. Dementavicius, D.; Lukseviciute, V.; Gómez-López, V.; Luksiene, Z. Application of mathematical models for bacterial inactivation curves using H ypericin-based photosensitization. *J. Appl. Microbiol.* **2016**, *120*, 1492–1500. [CrossRef]
67. Chang, W.; Small, D.A.; Toghrol, F.; Bentley, W.E. Global transcriptome analysis of Staphylococcus aureus response to hydrogen peroxide. *J. Bacteriol.* **2006**, *188*, 1648–1659. [CrossRef]
68. Charoux, C.M.G.; Patange, A.D.; Hinds, L.M.; Simpson, J.C.; O'Donnell, C.P.; Tiwari, B.K. Antimicrobial effects of airborne acoustic ultrasound and plasma activated water from cold and thermal plasma systems on biofilms. *Sci. Rep.* **2020**, *10*, 17297. [CrossRef]
69. Hozák, P.; Scholtz, V.; Khun, J.; Mertová, D.; Vaňková, E.; Julák, J. Further Contribution to the Chemistry of Plasma-Activated Water: Influence on Bacteria in Planktonic and Biofilm Forms. *Plasma Phys. Rep.* **2018**, *44*, 799–804. [CrossRef]
70. Liao, X.; Bai, Y.; Muhammad, A.I.; Liu, D.; Hu, Y.; Ding, T. The application of plasma-activated water combined with mild heat for the decontamination of Bacillus cereus spores in rice (*Oryza sativa* L. ssp. japonica). *J. Phys. D* **2019**, *53*, 064003. [CrossRef]
71. Krämer, C.E.M.; Wiechert, W.; Kohlheyer, D. Time-resolved, single-cell analysis of induced and programmed cell death via non-invasive propidium iodide and counterstain perfusion. *Sci. Rep.* **2016**, *6*, 32104. [CrossRef]
72. Kailas, L.; Terry, C.; Abbott, N.; Taylor, R.; Mullin, N.; Tzokov, S.B.; Todd, S.J.; Wallace, B.A.; Hobbs, J.K.; Moir, A.; et al. Surface architecture of endospores of the Bacillus *cereus/anthracis/thuringiensis* family at the subnanometer scale. *Proc. Natl. Acad. Sci. USA* **2011**, *108*, 16014–16019. [CrossRef]
73. Nicholson, W.L.; Munakata, N.; Horneck, G.; Melosh, H.J.; Setlow, P. Resistance of Bacillus endospores to extreme terrestrial and extraterrestrial environments. *Microbiol. Mol. Biol. Rev.* **2000**, *64*, 548–572. [CrossRef] [PubMed]
74. Russell, A.D. Bacterial spores and chemical sporicidal agents. *Clin. Microbiol. Rev.* **1990**, *3*, 99–119. [CrossRef] [PubMed]
75. Yu, H.; Chen, S.; Cao, P. Synergistic bactericidal effects and mechanisms of low intensity ultrasound and antibiotics against bacteria: A review. *Ultrason. Sonochem.* **2012**, *19*, 377–382. [CrossRef]
76. Zhang, R.; Ma, Y.; Wu, D.I.; Fan, L.; Bai, Y.; Xiang, Q. Synergistic Inactivation Mechanism of Combined Plasma-Activated Water and Mild Heat against Saccharomyces cerevisiae. *J. Food Prot.* **2020**, *83*, 1307–1314. [CrossRef]

77. Rosenberg, M.; Azevedo, N.F.; Ivask, A. Propidium iodide staining underestimates viability of adherent bacterial cells. *Sci. Rep.* **2019**, *9*, 1–12. [CrossRef]
78. Xiang, Q.; Zhang, R.; Fan, L.; Ma, Y.; Wu, D.; Li, K.; Bai, Y. Microbial inactivation and quality of grapes treated by plasma-activated water combined with mild heat. *LWT* **2020**, *126*, 109336. [CrossRef]
79. Fleet, G. Spoilage yeasts. *Crit. Rev. Biotechnol.* **1992**, *12*, 1–44. [CrossRef]
80. Herianto, S.; Hou, C.-Y.; Lin, C.-M.; Chen, H.-L. Nonthermal plasma-activated water: A comprehensive review of this new tool for enhanced food safety and quality. *Compr. Rev. Food Sci. Food Saf.* **2021**, *20*, 583–626.
81. Liao, X.; Su, Y.; Liu, D.; Chen, S.; Hu, Y.; Ye, X.; Wang, J.; Ding, T. Application of atmospheric cold plasma-activated water (PAW) ice for preservation of shrimps (Metapenaeus ensis). *Food Control* **2018**, *94*, 307–314. [CrossRef]
82. Vatansever, F.; de Melo, W.C.M.A.; Avci, P.; Vecchio, D.; Sadasivam, M.; Gupta, A.; Chandran, R.; Karimi, M.; Parizotto, N.A.; Yin, R.; et al. Antimicrobial strategies centered around reactive oxygen species–bactericidal antibiotics, photodynamic therapy, and beyond. *Fems Microbiol. Rev.* **2013**, *37*, 955–989. [CrossRef]
83. Imlay, J.A. The molecular mechanisms and physiological consequences of oxidative stress: Lessons from a model bacterium. *Nat. Rev. Microbiol.* **2013**, *11*, 443–454. [CrossRef]
84. Stadtman, E.R. Protein oxidation and aging. *Free Radic. Res.* **2006**, *40*, 1250–1258. [CrossRef] [PubMed]
85. Brioukhanov, A.; Netrusov, A. Aerotolerance of strictly anaerobic microorganisms and factors of defense against oxidative stress: A review. *Appl. Biochem. Microbiol.* **2007**, *43*, 567–582. [CrossRef]
86. Jemison, J.M., Jr.; Sexton, P.; Camire, M.E. Factors Influencing Consumer Preference of Fresh Potato Varieties in Maine. *Am. J. Potato Res.* **2008**, *85*, 140–149. [CrossRef]
87. Putnik, P.; Barba, F.J.; Lorenzo, J.M.; Gabrić, D.; Shpigelman, A.; Cravotto, G.; Bursać Kovačević, D. An integrated approach to mandarin processing: Food safety and nutritional quality, consumer preference, and nutrient bioaccessibility. *Compr. Rev. Food Sci. Food Saf.* **2017**, *16*, 1345–1358. [CrossRef]
88. Harker, F.R.; Gunson, F.A.; Jaeger, S.R. The case for fruit quality: An interpretive review of consumer attitudes, and preferences for apples. *Postharvest Biol. Technol.* **2003**, *28*, 333–347. [CrossRef]
89. Ma, R.; Yu, S.; Tian, Y.; Wang, K.; Sun, C.; Li, X.; Zhang, J.; Chen, K.; Fang, J. Effect of non-thermal plasma-activated water on fruit decay and quality in postharvest Chinese bayberries. *Food. Bioprocess Technol.* **2016**, *9*, 1825–1834. [CrossRef]
90. Carreño, J.; Martínez, A.; Almela, L.; Fernández-López, J.A. Proposal of an index for the objective evaluation of the colour of red table grapes. *Food Res. Int.* **1995**, *28*, 373–377. [CrossRef]
91. Huang, L.; Lu, J. The impact of package color and the nutrition content labels on the perception of food healthiness and purchase intention. *J. Food Prod. Mark.* **2016**, *22*, 191–218. [CrossRef]

MDPI

St. Alban-Anlage 66

4052 Basel

Switzerland

Tel. +41 61 683 77 34

Fax +41 61 302 89 18

www.mdpi.com

Foods Editorial Office

E-mail: foods@mdpi.com

www.mdpi.com/journal/foods

www.ingramcontent.com/pod-product-compliance
Lightning Source LLC
LaVergne TN
LVHW070640170726
843515LV00004B/290